D1300593

Oxford Physics Series

Oxford Physics Series

1. F. N. H. Robinson: *Electromagnetism*
2. G. Lancaster: *D.c. and a.c. circuits*
3. D. J. E. Ingram: *Radiation and quantum physics*
4. J. A. R. Griffith: *Interactions of particles*
5. B. R. Jennings and V. J. Morris: *Atoms in contact*
6. G. K. T. Conn: *Atoms and their structure*
7. R. L. F. Boyd: *Space physics; the study of plasmas in space*
8. J. L. Martin: *Basic quantum mechanics*
9. H. M. Rosenberg: *The solid state*
10. J. G. Taylor: *Special relativity*
11. M. Prutton: *Surface physics*

J. G. TAYLOR
KINGS COLLEGE, LONDON

Special relativity

Clarendon Press · Oxford · 1975

Oxford University Press, Ely House, London W.1

GLASGOW NEW YORK TORONTO MELBOURNE WELLINGTON
CAPE TOWN IBADAN NAIROBI DAR ES SALAAM LUSAKA ADDIS ABABA
DELHI BOMBAY CALCUTTA MADRAS KARACHI LAHORE DACCA
KUALA LUMPUR SINGAPORE HONG KONG TOKYO

CASEBOUND ISBN 0 19 851823 4

PAPERBACK ISBN 0 19 851824 2

PRINTED IN GREAT BRITAIN
BY J. W. ARROWSMITH LTD, BRISTOL

Editor's foreword

RELATIVITY is now a mature branch of physics, yet it continues to fire the interest of both science and mathematics undergraduates and experimental and theoretical physicists at the research level, as well as many non-scientists. The many books in this area aim at readership varying from the interested amateur to specialist sub-sects within the field. Professor Taylor's contribution to the Oxford Physics Series forms the basis of a first course in relativity for physics or mathematics undergraduate students.

The early chapters review special relativity from an elementary mathematical viewpoint, and include discussion of recent experiments which set out to test Einstein's predictions. The theory of relativity is then reformulated in more sophisticated mathematical language to show its relation to electromagnetism, and to lay the foundation for more general viewpoints. Here a knowledge of vector calculus and electromagnetism is needed to work through the details of the arguments, though a picture of the mathematical structure of relativity theory can be obtained with a very limited background in these areas. The final chapter discusses in simple terms where activity in the field is currently centred, and where future interest lies.

This text complements the Oxford Physics Series books on *Mechanics and motion* (*OPS 5*) and *Electromagnetism* (*OPS 1*), but gives a self-contained treatment of the subject.

J.A.D.M.

Preface

THE theory of special relativity is almost 70 years old. A great deal of scientific progress has been achieved in this theory since then. So much so that it might even be said that as far as special relativity is concerned all has been worked out and tested; at which stage, of course, a theory may rapidly become fossilized and lose its excitement and challenge. This is dangerous on two counts. First, being of such venerable age, special relativity is used extensively as an intrinsic part of the physicist's tool box. Having lost its intellectual excitement it may become more difficult to learn than would otherwise be the case. Secondly, any scientific theory is fallible, so that it should be taught in a manner in which its crucial successes and possible failings are made evident.

The present text arose from a course on the sources of applied mathematics given to first-year undergraduates at King's College, London. The course was designed to be taken in conjunction with other courses dealing with linear algebra, partial differentiation, and elementary vector field theory, so that the necessary mathematical tools were there when needed. The purpose of the course was to describe the development of special relativity as a theory based on experience. In particular, as little of classical mechanics as possible was assumed, at least in the earlier discussions of space and time distortion with motion. Throughout, the experimental basis of special relativity was emphasized, with its remarkable successes, and absolutely no failures.

The content of the course has been expanded slightly in the final chapters, especially with reference to the covariant formulation of electrodynamics and acceleration and its local equivalence with gravity. It was felt that these topics lead most naturally to remarks on the areas of greatest research interest in relativity. In particular the problems of black holes and of the quantization of gravity are now regarded as the two most crucial ones facing us at present in physics. It is to be hoped that the challenges they present are made more accessible to those who have already traversed the earlier stages of special relativity as presented in this book.

Finally, I would like to thank my students for their very honest responses to the course which helped to test it under fire, and to my typist, Pam Taylor, for her devotion to a difficult job.

J. G. TAYLOR

Contents

1. The constant speed of light

1.1. Introduction

THE speed of light in a vacuum has a constant value, independent of the velocity of the source of the light, of the observer, or of the nature of the light itself. Such constancy is contrary to everyday experience about relative motion and to its quantitative expression in Newton's laws of motion. The attempt at the turn of the last century and the beginning of this century to reconcile these two opposing results led to the creation of the theory of special relativity and the destruction of Newtonian mechanics. Since so much in this development hinges on the constancy of the speed of light, we will begin the study of special relativity with a brief description of the experiments validating this crucial fact, and an initial discussion of its importance.

By light is often meant visible light, though this, in truth, occupies only a very small portion of the electromagnetic spectrum. While nearly the most accurate determination of the speed of light has been for visible light, there is less than 1 per cent difference in the measured velocity of electromagnetic radiation over a range of wavelengths from the long radio waves down to the very energetic X-rays. The results of a number of experiments on this are shown in Table 1; the total spread of wavelengths is by a factor of 10^{15}. Microwave experiments are of very high accuracy, and though the wavelengths differ by a factor of 10^4 the speeds are the same to 1 part in 10^6.

TABLE 1

The speed of light at different wavelengths

Wavelength/m	Speed ($\times 10^8$) m s^{-1}
6·4	$2{\cdot}9978 \pm 0{\cdot}0003$
1·8	$2{\cdot}99795 \pm 0{\cdot}00003$
1·0	$2{\cdot}99792 \pm 0{\cdot}00002$
10^{-1}	$2{\cdot}99792 \pm 0{\cdot}00009$
$1{\cdot}2 \times 10^{-2}$	$2{\cdot}99792 \pm 0{\cdot}000003$
$4{\cdot}2 \times 10^{-3}$	$2{\cdot}997925 \pm 0{\cdot}000001$
$5{\cdot}6 \times 10^{-7}$	$2{\cdot}99793 \pm 0{\cdot}000003$
25×10^{-12}	$2{\cdot}983 \pm 0{\cdot}015$
$7{\cdot}3 \times 10^{-15}$	$2{\cdot}97 \pm 0{\cdot}03$

Similar agreement occurs with light from a moving source. For example, the light emitted in the decay of high-speed unstable particles which are travelling at about 99·97 per cent of the speed of light has a measured velocity in agreement with the values obtained for stationary sources to 1 part in 10^4, which is remarkably close.

1.2. Measuring the speed of light

The measurement of the speed of a moving object is usually achieved by determining the time the object takes to travel a given distance. Thus two measurements have to be performed; one of a time, one of a distance. Since light travels at the very great speed of about 186 000 miles, or 3×10^5 km, per second, it is necessary to observe the transit of light between very distance points, unless great accuracy in time-measurements can be achieved. This explains why the early attempt made by Galileo failed. He had two observers furnished with lamps and situated a distance apart. The first observer uncovered his lamp, and the second observer uncovered his as soon as he saw the light from the first observer's lamp. The distance between the lamps was too short for the first observer to determine the time taken for the light to travel from him to his colleague and back again.

That light had a finite velocity was not shown till 1675, when the Danish astronomer Romer noticed that the interval of time between successive eclipses of one of the four moons of Jupiter varies according to the distance of Jupiter from the earth. When the earth was approaching Jupiter the time interval was decreasing; as the earth receded so the interval increased. Romer explained this phenomenon by assuming that light took a finite time to travel through space. S represents the sun in Fig. 1 and E, E′ are corresponding

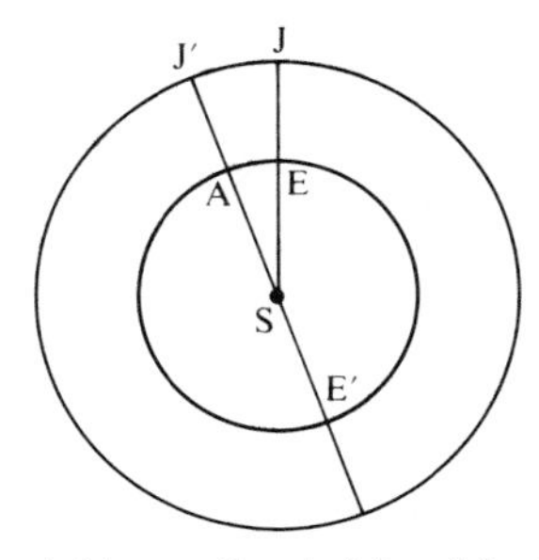

FIG. 1. The orbits of the earth (the smaller circle) and Jupiter (the larger one) around the sun S. The earth has travelled from E to E′ when Jupiter has moved from J to J′, the distance JE and J′E′ being of closest and farthest approach between the two planets.

positions of the earth when Jupiter is nearest to it (at J) or furtherest from it (at J′) (Jupiter takes 11·86 years to make a revolution round the sun). We can consider the eclipses of a Jovian moon as providing a regular sequence of time signals. These are received at a successively later time as the earth moves from E to E′; their maximum delay will be the time taken for light to travel from A to E′. Thus the velocity of light will be

$$c = \frac{\mathrm{AE}'}{\Delta t}.$$

The measurement of AE′, the diameter of the earth's orbit, is obtained from the solar parallax, the angle subtended by the earth's radius at the sun. Using a recent value of 8·79″ for the latter and 1002 s for Δt gives the result

$$c = 2{\cdot}98 \times 10^5 \text{ km s}^{-1}.$$

This method has limited accuracy, so we will turn to more accurate ones.

The most precise measurements of the speed of visible light used a source varying with an accurately known frequency ν. A suitable beam of this light was reflected from a mirror M (see Fig. 2) and picked up by a photo-cell C

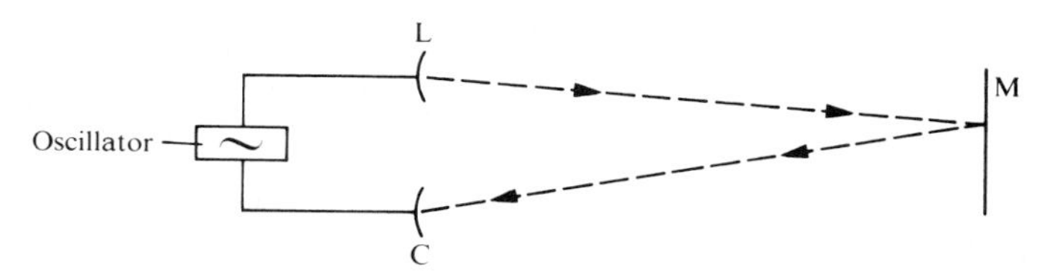

FIG. 2. Apparatus to measure the velocity of visible light. The oscillating source sends light to the mirror M which is reflected to the photo-cell C powered by an identically oscillating energy supply.

whose sensitivity varied at the same frequency ν. Certain positions of the mirror produced a 'resonance' condition when the flashes from L reached the tube at instants of high sensitivity. If the mirror had to be moved a distance d to go from one resonance position to the next then it took the light a single period to travel this extra distance, so

$$c = 2d\nu$$

(where allowance has still to be taken for divergence of the beam from the vertical on reflection at M). Using suitable frequencies in the megacycle range allowed an accuracy of 1 part in 10^7 to be obtained, as is evident from Table 1.

A method with even greater accuracy used microwave radiation, especially in the millimetre wavelength range. The microwave beam was split into two parts, one of which travelled over a fixed path, the other over a variable one. These two beams were combined at a detector. When the path difference between the two beams was a whole number of wavelengths the two beams

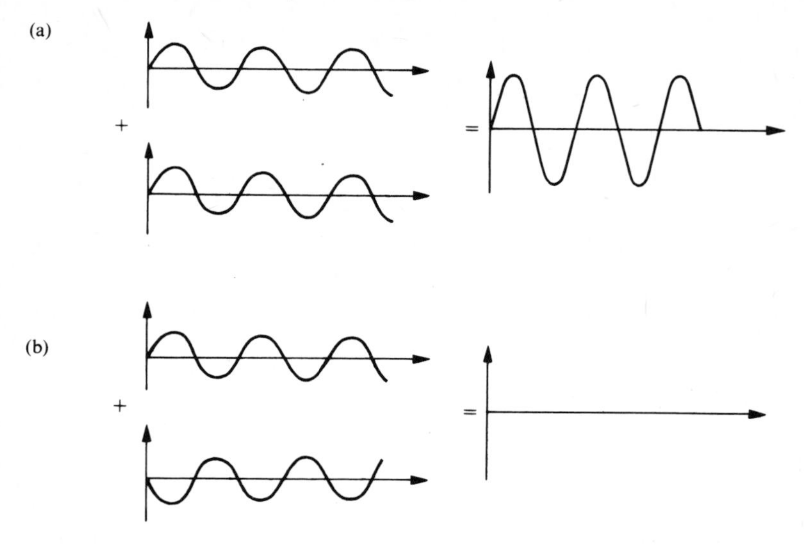

FIG. 3. The nature of constructive or destructive interference. In (a) the two waves are in phase, so add, while in (b) phase difference is one half a wavelength and they destroy each other.

added to give a maximum effect; when the path difference was half an odd integer number of wavelengths the beams destroyed each other and produced a minimum effect, as shown in Fig. 3. Thus the distance needed to move the detector so as to go from a position of maximum intensity at the detector to a successive one was one wavelength λ, and was measured with considerable accuracy. The frequency ν of the microwave radiation was also carefully determined, so allowing c to be obtained from

$$c = \nu\lambda.$$

There are numerous other methods of measuring c, though none with such accuracy; they all give results in agreement with each other, as is clear from Table 1.

1.3. The nature of light

In order to understand the constancy of the velocity of light we have to consider the nature of light. There have been two opposing theories as to how light energy is transported from one point to another. The view most simply explaining its rectilinear propagation is that light consists of a stream of particles emitted by the source. This was proposed originally by

Pythagoras in the sixth century B.C., but met great difficulty in the seventeenth century when the phenomena of interference and diffraction were discovered—that brilliant colours were produced by a thin film of air between two glass surfaces or that light encroached upon the region of the geometrical shadow. This led to the opposing theory that light is a wave motion, for only then could these interference and diffraction effects be sensibly explained.

A further result which destroyed the particle theory of light was its constant speed independent of its source; this has been mentioned already in § 1.1. There has been absolutely no contradiction to this result for speeds of the source lower than the highest recorded one of 99·975 per cent of the speed of light. The difficulty which this presents to the particle theory of light is as follows.

Suppose we are observing in our laboratory light travelling with velocity c. If somebody passes by in a laboratory travelling with velocity v we might compare notes with him about our relative observations of common phenomena, such as the speed of the ray of light we have just set up. Our comparisons would be initially between events labelled with coordinates x, y, and z in our laboratory and x', y', and z' in our colleague's. If we suppose that he is moving along the direction of our x-axis and that the origins of coordinates coincide at time $t = 0$, we relate coordinate labels in the two laboratories by

$$x' = x - vt, \qquad y' = y, \qquad z' = z. \tag{1.1}$$

But then the velocity of the ray of light which we have measured to be c should be $(c-v)$ according to our colleague. In general, velocities are related by

$$\dot{x}' = \dot{x} - v, \qquad \dot{y}' = \dot{y}, \qquad \dot{z}' = \dot{z}, \tag{1.2}$$

where the dot denotes differentiation with respect to the universal time t,

$$\dot{a} = \frac{\mathrm{d}a}{\mathrm{d}t}.$$

The transformation equations (1.1) and (1.2) are called Galilean, in honour of Galileo who first considered the manner in which velocities should be related by relatively moving observers.

The constant speed of light would seem to be at variance with the Galilean transformation laws (1.1) and (1.2), which are at the heart of Newtonian mechanics. In the latter it is only acceleration which enters meaningfully into Newton's second law of motion,

$$\text{force} = \text{mass} \times \text{acceleration}$$

or

$$\boldsymbol{F} = m\ddot{\boldsymbol{r}}, \tag{1.4}$$

with vector notation $\boldsymbol{F} = (F_x, F_y, F_z)$, $\boldsymbol{r} = (x, y, z)$. From (1.1) and (1.2),

$$\ddot{\boldsymbol{r}}' = \ddot{\boldsymbol{r}}, \tag{1.5}$$

so that acceleration is an invariant. If the force $\boldsymbol{F}$ be unchanged also in moving from a given laboratory to one moving at constant speed with respect to it, then Newton's second law will be unmodified. Forces that depend only on the relative positions or velocities of the objects between which they act will certainly have this property, since then if positions of interacting objects in the two reference frames are $\boldsymbol{r}_1, \boldsymbol{r}_2, \ldots$; $\boldsymbol{r}'_1, \boldsymbol{r}'_2, \ldots$, we have from (1.1) and (1.2)

$$\boldsymbol{r}'_1 - \boldsymbol{r}'_2 = \boldsymbol{r}_1 - \boldsymbol{r}_2,$$

$$\dot{\boldsymbol{r}}'_1 - \dot{\boldsymbol{r}}'_2 = \dot{\boldsymbol{r}}_1 - \dot{\boldsymbol{r}}_2.$$

Thus forces depending only on the quantities $\boldsymbol{r}_1 - \boldsymbol{r}_2$, etc., and $\dot{\boldsymbol{r}}_1 - \dot{\boldsymbol{r}}_2$, etc. will be unchanged in the new coordinates.

Newton's laws proved of enormous success for a period of over two centuries and are still eminently satisfactory to describe an enormous range of macroscopic phenomena. Yet the propagation of light, if it always occurs at the same speed, cannot be described in a fashion consistent with these laws; it cannot satisfy the Galilean transformation law (1.2). And certainly if light does consist of particles they cannot be described by Newton's laws. But then the attraction of the particulate theory of light is lost, since there is no need for these particles to travel in straight lines; their behaviour becomes totally mysterious.

On the other hand, the wave theory of light required the existence of a medium through which light vibrated, analogous to transverse waves on a string. This medium was called the luminiferous ether and was required to have very strong restoring forces so that it could propagate light at such a great speed. But at the same time the medium could offer little resistance to the planets, since they suffered no observable reduction in speed even though they travelled through it for year after year. It was necessary to demonstrate the existence of the ether so that this paradox might be resolved.

1.4. The search for the ether

The first clear property of the ether was that it was only little dragged along by the earth in its motion through the heavens. This was apparent from the phenomenon of aberration, first discovered by the British astronomer Bradley in 1725. He found that the altitude of a star varied with the position of the earth in its orbit round the sun, there being the greatest apparent stellar displacement when the earth was moving directly away from or towards the star.

The phenomenon of aberration can be explained in terms of the angle δ needed to tilt a telescope pointed originally at the star if the earth was stationary with respect to it, in order once more to have the star in the centre of the field of view when the earth is moving with velocity v. From Fig. 4(b) we see that

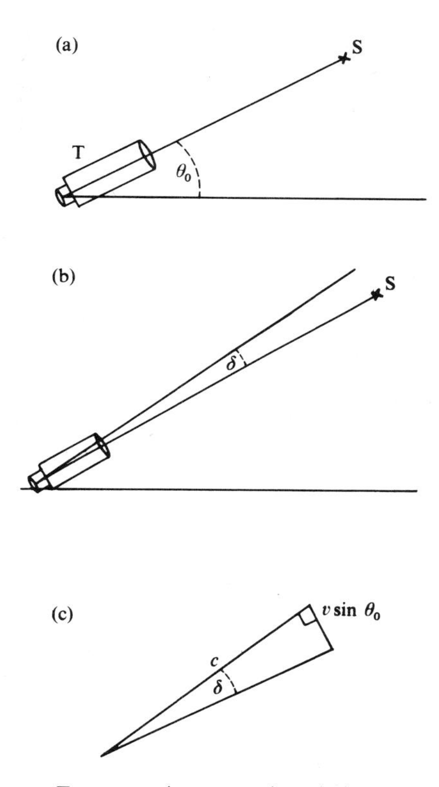

FIG. 4. (a) A telescope T on a stationary earth, pointing at a star of altitude θ_0. (b) The earth is now in motion with velocity v. (c) The velocity diagram for light entering the telescope in (b). It has a component perpendicular to the original position of T of value $v \sin \theta_0$ and a component c along the original direction of T, so that T must be tilted by $\delta \sim v \sin \theta_0/c$ so as to be parallel to the resultant direction of motion.

the aberration angle δ is of magnitude $v \sin \theta_0/c$. Experimentally, it oscillates with a maximum of about 40″ of arc for the star γ-Draconis, as is expected from a more complete analysis. This agrees very well with the value expected from the magnitudes of the orbital speed of the earth of about 30 km s^{-1} and of $\theta_0 = 75°$.

The simplest explanation of aberration is if light is particulate, for then the light motion just obeys the velocity transformation law (1.2). We have had to reject this model for light, and can only understand the effect in terms of waves if the ether remains almost completely unaffected by the earth's motion. Any appreciable dragging of the ether near the earth would prevent aberration from taking place.

In principle, it should be possible to detect the motion of the earth through the ether, if the latter exists and is not dragged along with the earth. The most sensitive search for this relative motion at the end of the last century was performed by Michelson and Morley, using the apparatus shown in Fig. 5.

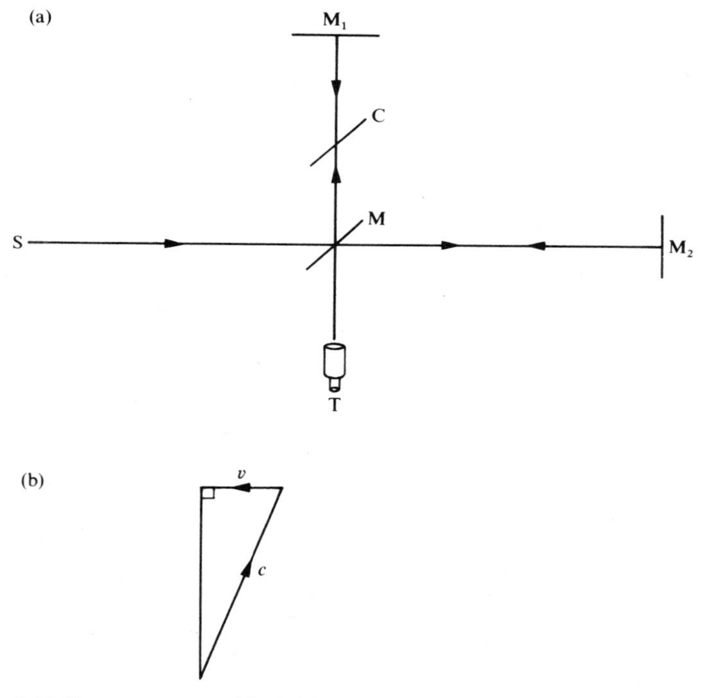

FIG. 5. (a) The apparatus used by Michelson and Morley to detect the motion of the earth through the ether. Light from a source S is split into two parts by the mirror M half-silvered on its front surface. M_1 and M_2 are two mirrors almost at 90° so that interference fringes are produced in the telescope T when the reflected beams recombine. C is a compensating sheet of glass. (b) The velocity diagram for light travelling along the path MM_1, being the resultant of v horizontally due to the motion through the ether and c in the ether at an appropriate direction so that the resultant is along MM_1.

When a beam of light from the source S fell onto the inclined glass plate M it was split into two portions, owing to a semi-transparent coating of metal on its front face. One half of the beam was transmitted, reflected from the mirror M_1, and a part of it reflected by M into the telescope T. The other half of the beam arising from reflection at M was reflected back to M by the mirror M_2, and a portion of it transmitted to T. A further plate C of glass was put in the path of this second part of the beam so that both parts travelled through the same thickness of glass.

If M is at an angle of 45° to the incoming beam of light and the surfaces of the mirrors M_1 and M_2 are very nearly, but not exactly, at 90° to each other, then interference fringes are formed. They resemble those formed by a wedge of very small angle, and are obtained due to the slight difference in path lengths travelled by the two parts of the beam (see Fig. 3 for conditions for the constructive or destructive interference of two waves). If the path lengths MM_1 and MM_2 are l_1 and l_2, which gives rise to a particular fringe, then

$$2(l_1 - l_2) = n\lambda,$$

where n is an integer and λ is the wavelength of the light used.

Let us suppose that the apparatus was moving with speed v along the direction MM_2 with respect to the ether. As viewed from the apparatus there was expected to be, on the ether hypothesis, an ether wind blowing past the apparatus, from M_2 to M, of speed v. This would have altered the velocity of the light travelling along MM_2 or MM_1 by calculable amounts. We must remember here that light was supposed to be moving at speed c with respect to the ether only, and so obeyed the usual Galilean composition law of velocities (1.2) when observed with apparatus moving through the ether.

The velocity of light along MM_2 was thus $(c-v)$, and along M_2M was $(c+v)$. The time for the light to travel from M to M_2 and back was then

$$t_2 = l_2\left(\frac{1}{c+v}+\frac{1}{c-v}\right) = \frac{2l_2}{c}\left(1-\frac{v^2}{c^2}\right)^{-1}. \tag{1.6}$$

The magnitude of the velocity of light along MM_1 should have been $(c^2-v^2)^{\frac{1}{2}}$, being the resultant of the light velocity c in a certain direction and the ether velocity v in a direction at right angles to MM_1 (as shown in Fig. 5(b)). The time taken for light to travel from M to M_1 and back was thus

$$t_1 = \frac{2l_1}{c}\left(1-\frac{v^2}{c^2}\right)^{-\frac{1}{2}}. \tag{1.7}$$

The time difference between the two is

$$\Delta = t_2 - t_1 \approx \frac{2(l_2-l_1)}{c}+\frac{2l_2v^2}{c^3}-\frac{l_1v^2}{c^3}, \tag{1.8}$$

when $v/c \ll 1$, so that the denominators in (1.6) and (1.7) can be expanded in powers of v/c. The whole apparatus was rotated through 90°, the new time difference being obtained from Δ by interchanging l_1 and l_2 in (1.8), so being

$$\Delta' \approx \frac{2(l_1-l_2)}{c}+\frac{2l_1v^2}{c^3}-\frac{l_2v^2}{c^3}. \tag{1.9}$$

Finally,

$$\Delta-\Delta' \approx \frac{4(l_2-l_1)}{c}+\frac{(l_1+l_2)v^2}{c^3}. \tag{1.10}$$

This change of time difference on rotation of the apparatus through 90° should have caused a shift in the interference fringes at the telescope T. Since c/λ is the frequency of the light, the number of waves added or removed by the 90° rotation was expected to be

$$n = \frac{c(\Delta-\Delta')}{\lambda}. \tag{1.11}$$

The value of n will thus be the number of interference fringes which were shifted past a given point by the rotation of the apparatus; this number (which may not be an integer) is, from (1.10) and (1.11),

$$n \approx \frac{4(l_2-l_1)}{\lambda}+\left(\frac{v}{c}\right)^2\frac{(l_1+l_2)}{\lambda}. \tag{1.12}$$

If we take $l_1 = l_2$ then the shift of the fringe pattern will be proportional to $(v/c)^2$, which will be rather small. In spite of this it proved possible to determine a relatively small upper bound. The original experiment detected a shift of at most 0·005 of a fringe; the value expected if v were the velocity of the earth in orbit (about 30 km/s^{-1}) is 0·4 of a fringe. The experiment has been performed subsequently using two lasers operating at almost identical frequencies of about 3×10^{14} Hz in the positions of the mirrors M_1 and M_2. No change of frequency difference was observed between the two lasers on rotation of the apparatus through 90°; the velocity of the earth relative to the ether was less than one thousandth of that of the earth in orbit, and showed no indication of any movement of the earth through the ether.

1.5. The absence of the ether

The phenomenon of aberration indicated that if a luminiferous ether existed then it was being dragged along with the earth. But the Michelson–Morley experiment appeared to contradict this. To save the situation it was suggested that there is a contraction of lengths in the direction of motion by the factor $(1-v^2/c^2)^{\frac{1}{2}}$. For then the length is reduced by this factor, so that

$$\frac{2(l_2-l_1)}{c}\left(1-\frac{v^2}{c^2}\right)^{-\frac{1}{2}} \tag{1.13}$$

and will vanish if $l_1 = l_2$. Similarly, after rotation through 90° the length l_1 will suffer reduction by the factor $(1-v^2/c^2)^{\frac{1}{2}}$, so

$$\Delta' = \frac{2(l_1-l_2)}{c}\left(1-\frac{v^2}{c^2}\right)^{-\frac{1}{2}}. \tag{1.14}$$

It will also vanish if $l_1 = l_2$, so that there will be no time difference on rotation of the apparatus.

This explanation of the result of Michelson and Morley is still not enough, however. A modification of the apparatus was made so that the arms of the interferometer were of different length, and furthermore the apparatus was fixed in the laboratory and observed over a period of months. Thus the time taken for light to travel the extra distance $2|l_2-l_1|$ must be independent of the velocity of the laboratory, which contradicts the existence of an ether through which light propagates at the speed c.

Nor will the space contraction introduced above remove this difficulty and still allow the ether to perform its necessary function. For we see that even with space contraction, $\Delta-\Delta' = \{4(l_2-l_1)/c\}(1-v^2/c^2)^{-\frac{1}{2}} \neq 0$. Nor do we expect the time difference between the paths to vanish at other orientations of the apparatus to the direction of motion of the ether wind.

At this point we see that a more complete revision of the nature of space, and even of time, are necessary. In particular the change of coordinates described by (1.1) needs to be considered afresh. There is already evidence that distances may suffer a contraction when viewed from a moving frame. Even time itself may have to be modified. We must attempt to do this, however, only on the basis of the crucial experimental results described in this chapter; the velocity of light is independent of the velocity of its source, of the observer, and of the nature of the light itself.

Further reading for Chapter 1

A useful and relatively complete discussion of measurements of the velocity of light is given by J. H. Sanders in *The velocity of light*, Pergamon Press, Oxford (1965).

Earlier measurements are also clearly discussed in J. F. Mulligan's, Some recent determinations of the velocity of light, *Am. J. Phys.* **20**, 165–72 (1972) and by J. F. Mulligan and D. F. McDonald in Some recent determinations of the velocity of light II, *Am. J. Phys.* **25**, 180–92 (1957).

More recent measurements are critically discussed in the paper, Determination of e/h, Q.E.D., and the fundamental constants, by B. N. Taylor, W. H. Parker, and D. N. Langenberg, *Rev. mod. Phys.* **41**, 385–7 (1969).

A useful survey of the Michelson–Morley experiment is contained in B. Jaffe's *Michelson and the speed of light*, Science Study Series no. 11, Heinemann, London (1961).

2. Measuring time and distance

2.1. Changing space and time

LIGHT travels at constant speed, but we saw in the previous chapter that there is no mechanical model, such as the luminiferous ether, which is consistent with the various features of light propagation. The existence of an unvarying velocity for light is in disagreement with Newton's laws of motion, and even more generally with the Galilean laws of transformation of space and time as described by eqns (1.1) and (1.2).

It was pointed out earlier that a partial understanding of the Michelson–Morley experiment could be obtained if there was a reduction of length along the direction of motion of the apparatus. In other words, a metre rod would appear shorter by a factor $(1-v^2/c^2)^{\frac{1}{2}}$ when viewed by someone moving past it with speed v. This space contraction was still not satisfactory to explain the null result of the extended Michelson–Morley experiment, in which the two arms of the interferometer had unequal length. It was noted that a distortion of time might allow us to comprehend the further lack of evidence of the existence of the ether, but that would be a radical step indeed. But it is all that is available; all possible modification of space has been achieved by the contraction factor $(1-v^2/c^2)^{\frac{1}{2}}$.

To require time to be altered in any fashion was truly unthinkable before the beginning of this century. To Newton: 'Absolute, true and mathematical time, of itself, and from its own nature, flows equably without relation to anything external.' Time does, indeed, seem to possess this objective property; a watch or clock continues its inexorable ticking at a rate which seems to be completely independent of how fast it is travelling or how it is observed.

It is necessary to reject the notion of a universal time before the phenomenon of the constant velocity of light can be understood. Evidently a careful analysis of time measurement is necessary to justify such a big step. We will have to carefully consider the way in which time may be measured when using light; we will arrive at the phenomenon of time dilation: 'moving clocks go slow'. This is still only an initial step in incorporating light propagation into the fabric of physics. The space-contraction factor will then be obtained from a further examination of how distances are measured. This analysis will allow us a pragmatic basis on which to build the edifice of special relativity in which space and time modifications are combined together in the four-dimensional space–time continuum.

2.2. Time dilation

We will assume that all clocks are alike in that their rates, if modified in any way by motion, are all altered by the same amount. If this were not the case then it would be possible to detect an absolute velocity as that for which there was, say, least variation among the rates of a chosen set of different clocks moving at that speed. This is certainly at variance with the principle of relativity, that one laboratory, assumed proved with a set of measuring rods and clocks, appears no different to experimenters working in it than another laboratory moving uniformly with respect to the first would to its complement of researchers. We will have more to say about relativity later, but since there is no evidence at all for the existence of such absolute velocities we will assume that the principle of relativity in this form is valid.

It is useful to note here that a laboratory such as the one mentioned above, provided with a set of measuring rods and clocks, is called a frame of reference. The experimentalists associated with this frame are allowed to travel unhindered throughout space and to make measurements of distances and times when at rest with respect to the laboratory. It is even possible to envisage a frame as provided with a set of observers located at many points throughout space, all moving at the same speed as the laboratory and each furnished with a clock and a metre rule. The clocks are assumed to be synchronized, and all keep perfect time. The concepts of a frame of reference and an observer attached to such a frame will be used constantly throughout the text, so it is advisable that the reader ensure that he or she is familiar with them.

We will restrict our discussion to inertial frames of reference, so called because in them the laws of inertia—any free body continues in rectilinear motion—is satisfied. Mechanical experiments performed in an inertial frame are not expected to reveal its state of motion. If the velocity of light is the same in all inertial frames then neither will optical experiments allow such a motion to be discovered. All inertial frames of reference are therefore to be regarded as equivalent as far as scientific laws are concerned. This is usually called the principle of Special Relativity. Such equivalence of dynamics is evidently not the case when acceleration is considered, since extra effects depending crucially on the acceleration, such as centrifugal and Coriolis forces, occur. We will only consider how acceleration is to be included at a much later stage.

In order to incorporate the constancy of the velocity of light in our discussion of time-measurement we will consider a clock whose time-keeping properties are based directly on the propagation of light. The simplest form of such a clock is the light-pulse clock shown in Fig. 6(a). It is composed of two mirrors M_1 and M_2 held facing each other at the opposite ends of a rigid cylinder. The distance between M_1 and M_2 is taken to be l_0. A pulse of light bounces back and forth from one mirror to the other. There is a dial which records the return of the pulse at one end, so allowing the passage of time to be recorded. One 'tick' of the clock is the time a light pulse takes to make the round trip from one mirror to the other, be reflected, and so return from whence it came.

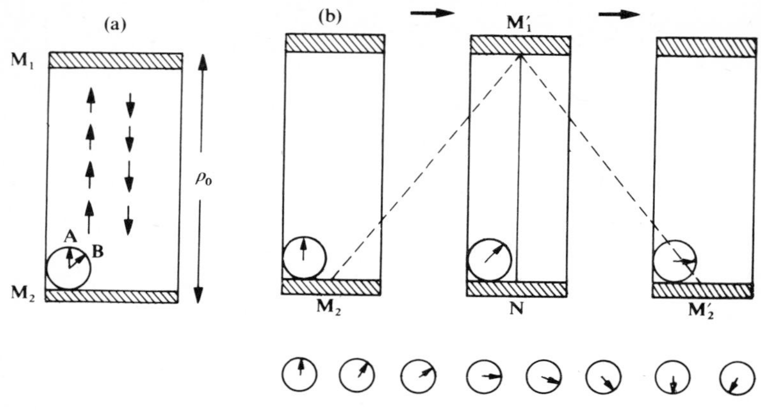

FIG. 6. The light-pulse clock. (a) When at rest a light pulse is reflected back and forth from the mirrors M_1 and M_2; each return of the pulse to one end counts as a unit of time. The dial hand moves from A to B in that time. (b) When the clock is in motion the light pulse travels the distance $M_2M_1'M_2'$ between ticks, so causing it to go slow. The dials underneath the moving clock are those stationary clocks distributed in space and used for comparison with the moving clock times.

Let us consider the clock moving to the right with velocity v. The path of a particular light pulse, as seen in a stationary frame of reference, is then $M_2M_1'M_2'$, as shown in Fig. 6(b). We suppose that the time taken for the motion from M_2 to M_2' is t, as measured by a stationary clock. We assume that there are a large number of these stationary clocks, all synchronized with each other and provided with observers, distributed throughout space, and so forming a frame of reference as described earlier. This allows a comparison to be made of the moving and stationary clock times by observing simultaneously the moving clock reading and that of the nearby stationary clock. The distance M_2M_2' is equal to vt and M_2N is $\frac{1}{2}vt$. Then

$$M_2N = \tfrac{1}{2}vt, \qquad M_1'N = l_0,$$

so

$$M_2M_1' = (l_0^2 + \tfrac{1}{4}v^2t^2)^{\frac{1}{2}},$$

and the distance travelled by the light pulse, as observed in the stationary frame, is

$$2M_2M_2' = 2(l_0^2 + \tfrac{1}{4}v^2t^2)^{\frac{1}{2}}.$$

The light pulse is travelling with speed c for a time t in this stationary frame.

Thus it travels a distance equal to ct as seen by a stationary observer, and so

$$ct = 2(l_0^2 + \tfrac{1}{4}v^2t^2)^{\frac{1}{2}}. \tag{2.1}$$

Solving (2.1) for t, we obtain

$$t = \frac{2l_0}{c}\left(1 - \frac{v^2}{c^2}\right)^{-\frac{1}{2}}. \tag{2.2}$$

But $2l_0$ is the distance travelled by the light pulse from the mirror M_2 back again, as seen by an observer moving with the clock. The light pulse travels with the same speed c as before, but now with regard to the frame moving with the clock (light *always* travels with velocity c, as we have seen in the previous chapter). The time t_0 measured by the moving clock for the light pulse to move back and forth is thus

$$t_0 = \frac{2l_0}{c}. \tag{2.3}$$

Combining (2.2) and (2.3) we obtain

$$t = t_0\left(1 - \frac{v^2}{c^2}\right)^{-\frac{1}{2}}. \tag{2.4}$$

For $v < c$ formula (2.4) implies that $t > t_0$. In other words, the time between the beats of a moving light-pulse clock, as measured by a stationary clock, is longer than that determined when moving with the clock. That is, 'moving clocks go slow' by an amount given by the time-dilation formula (2.4).

We have obtained a time-distortion formula which destroys Newton's idea of a universal time. The rate at which a clock advances depends on how fast one is travelling with respect to it, and is given by eqn (2.4). Such a modification of time is evidently so important that we must analyse it further before we can accept it. The most crucial thing to be done is to put time dilation to experimental test. If it satisfies all the experiments we can think of then we must attempt to incorporate it more fully into the laws of physics.

Before we do that let us clear up an apparent difficulty raised by (2.4); one that has been claimed to make the discussion self-contradictory. If we observe the moving light-pulse clock we see that it should go slow according to (2.4). But an observer moving with the clock would observe our own clock to go slow as compared to his, also by the same factor (2.4).

How is it possible that two clocks can both go slow with respect to each other? Indeed they cannot, but in any case the paradox has been caused by an imprecise use of words. A more correct phrase to use is 'moving clocks appear to go slow', so that while the accounts by the two observers of the rates of each other's clock is contradictory, neither need be lying. They can, indeed, both see each other's clock going slower than their own. The accounts they give do not need to tally. For the clocks are not going slow with respect to each other *in an absolute sense* but only in a relative one—relative, that is, to the particular observer who is moving with respect to them. To be quite precise one needs to rephrase (2.4) so as to read:

'moving clocks go slow relative to those of stationary observers'.

This form of the statement of time dilation indicates the truly relative nature of time and prevents the paradox mentioned above from ever arising.

There is a further feature of (2.4) which should be remarked upon here. The time-dilation factor $(1-v^2/c^2)^{-\frac{1}{2}}$ is not defined at $v = c$, while for $v > c$ it becomes imaginary. Since it is not possible to measure an imaginary time it would seem that c is a limiting velocity. A higher speed could only appear to be attained if the formalism breaks down. We will accept for the moment the implication that all velocities are limited by that of light. It is a feature that has recently been questioned, but we will only do so when we have reached a position where we can see more clearly the various implications of faster-than-light travel. In any case, we will see that we should not expect particles to be accelerated through the speed of light; once they travel slower than light we will find that they will always have to do so.

2.3. Measuring time dilation

The time-dilation factor $(1-v^2/c^2)^{-\frac{1}{2}}$ is so close to unity as to give very little chance of measuring it in normal everday experience. For example, a jet plane travelling at 600 m.p.h. has a value of v/c only a little over 10^{-6}, and the time-dilation factor is unity to within 5×10^{-10} per cent. Even a satellite travelling at 5 m.p.s. only has v/c about 3×10^{-5}, so a time-dilation factor only closer to unity than for the jet by a factor of 10^3.

In order to obtain values of v closer to c it is necessary to look at the sub-nuclear particles which can be made to travel at very close to the speed of light since they possess so little mass. The right candidate for further investigation would be a fast-moving particle which carries its own natural clock. A time-piece of this sort would be provided if the particle decayed in a certain lifetime which was long enough to be measured.

A very suitable particle is the mu-meson or muon, for short, denoted by μ. This is a particle which is about 200 times heavier than an electron but with identical electric charge. In fact the muon comes with either a negative or positive electric charge denoted μ^- or μ^+, in similar manner the electron is denoted by e^- and its oppositely charged but otherwise identical partner the positron is written e^+.

The muon is an unstable particle, decaying into three others, two of which are neutrinos, which are also similar to electrons except that they possess no electric charge or mass. The decay

$$\mu^\pm \to e^\pm + 2\nu$$

takes place with a lifetime τ of $2{\cdot}2\times10^{-6}$ s for muons at rest. This means that in a sample of muons e^{-1} of them will have decayed in τ s, and the fraction left after time t will be $e^{-t/\tau}$.

Muons are produced in the earth's atmosphere by cosmic rays coming from outer space; they can be detected by suitable charged-particle detectors. The muons created in this fashion have a very high velocity v, so they provide a fast moving clock with a natural 'tick' of τ s. It is possible to observe the number of them arriving at the top of a mountain, and also to determine the number which survive the trip down to sea-level. This can be compared with the number which would have been expected to survive if the muons had decayed with the lifetime τ. A discrepancy between these two values should be satisfactorily accounted for by the time-dilation factor (2.4).

We note that the phenomenon of time dilation is essential even to explain the overall feature that cosmic ray muons are observed at sea-level yet are made at an altitude of about 10 km. If they decayed in the lifetime τ equal to that at rest then they would only be able to travel a distance of 600 m in that time. They are observed at 17 times that distance, by which time there should be no trace of any muons surviving. Time dilation is essential to explain this phenomenon.

TABLE 2

Muons decaying at rest in a particular experiment

Elapsed time/μ s	0	1	2	3	4	5	6	7	8
Number of muons surviving	568	373	229	145	99	62	36	17	6

In Table 2 is shown the number of stationary muons which survive after a certain length of time in a particular experiment. We have to compare this with the fact that about 563 muons arrived per hour at counting apparatus set up at an altitude of 2000 m, while 400 muons were found each hour at sea-level as surviving from these high-altitude particles.

Let us suppose that the muons were created at high altitude so as to have a velocity very close to that of light (we are accepting the restriction that light is the limiting velocity, as we remarked at the end of the last section). They would then take about 6·5 μs to travel down to sea-level. From Table 2 we would only expect about 25 muons per hour to survive. But 400 muons per hour actually arrive, this being the number expected to survive after 0·7 μs, according to Table 2 (suitably interpolated). In other words, the high-speed muons themselves experience a time of 0·7 μs while the stationary clocks have ticked through 6·5 μs.

This discrepancy between the rates of the clocks can only be removed by means of the time-dilation factor (2.4). The time t is that of the stationary clocks and t_0 that experienced by the high-speed muons. The ratio of the times t/t_0 in that formula is

$$\frac{t}{t_0} = \frac{6{\cdot}5}{0{\cdot}7} \sim 9,$$

so that from (2.4)

$$\left(1 - \frac{v^2}{c^2}\right)^{-\frac{1}{2}} = 9$$

or

$$\frac{v}{c} \sim 0{\cdot}994. \tag{2.5}$$

We see that the muons are indeed moving very close to the speed of light; if we had used the value of v given by (2.5) to measure the time taken by the muons to reach sea-level there would have been almost no change in the final value of v resulting from the use of (2.4).

This test used muons for which the time-dilation factor was large. More recently, the factor has been shown to occur even at the speed of a commercial jet plane of about 600 m.p.h. The time-dilation factor was remarked on earlier as being very close to unity, so time changes due to its presence can only be determined with utmost precision. It was possible to achieve the required accuracy by using four cesium clocks and taking their averaged time. The four clocks were flown by commercial jet planes on two trips round the globe, one eastward, the other westward.

On the eastward trip there was a time loss of about 59 ns (1 ns $= 10^{-9}$ s), while westward motion produced a time gain of about 273 ns. These were all observed in relation to standard cesium clocks at the U.S. Naval Observatory in Washington, U.S.A.

That there was a time *gain* on the westward trip arises from the rotation of the earth. If it is rotating with angular velocity Ω and has radius R then a clock on its surface at the equator is also moving now with speed $R\Omega$ with respect to a hypothetical clock of an underlying nonrotating inertial frame. It therefore has a time-dilation factor which has to be taken account of, the factor being equal to

$$\left(1 - \frac{R^2\Omega^2}{c^2}\right)^{-\frac{1}{2}} \sim 1 + \frac{R^2\Omega^2}{2c^2}. \tag{2.6}$$

and where we neglect higher powers of $R\Omega/c$ above the second since

$$\frac{R\Omega}{c} \sim 1{\cdot}7 \times 10^{-6}.$$

A clock moving with speed v relative to the earth in an eastward direction has speed $(v+R\Omega)$, and so time-dilation factor is

$$1+\frac{(R\Omega+v)^2}{2c^2}.$$

The relative time-dilation factor between the clocks in the jet plane and those fixed on the earth is thus

$$1+\frac{(2R\Omega v+v^2)}{2c^2}. \tag{2.7}$$

and the eastward-travelling clocks should lose time compared to the stationary ones. The westward-moving clocks will have a time-dilation factor

$$1+\frac{(\Omega R-v)^2}{2c^2} \tag{2.8}$$

since they are now travelling in a direction opposite to the earth's rotation. The relative dilation factor obtained from (2.8) and (2.6), is

$$1-\frac{R\Omega v}{c^2}+\frac{v^2}{2c^2}. \tag{2.9}$$

Since the third term in (2.9) is at most a quarter of the second term for a commercial jet travelling at 600 m.p.h., the motion in a westward direction produces a gain in the travelling clocks relative to those stationary on the earth.

The values expected from the appropriate time-dilation factor, together with an altitude factor arising from gravitational effects are shown in Table 3.

TABLE 3

Time differences predicted and observed for clocks travelling by jet round the earth

Mean time difference of four cesium clocks	$\Delta\tau$/ns	
	Eastward	Westward
Observed	-59 ± 10	273 ± 17
Predicted	-40 ± 23	275 ± 21

The agreement between theory and experiment is remarkably good, and leaves no doubt as to the fact that moving clocks do go slow, even on jet planes. Moreover it shows that any effects of the slight acceleration at the beginning and end of the trip are negligible, as was to be expected.

2.4. Space contraction

We turn now to investigate how lengths are affected when we incorporate the constancy of the speed of light into the measurement process. We can again do that by means of the light-pulse clock of §2.2, though now with it moving in a direction parallel to its length. Let us suppose the velocity of this motion is v. We show in Fig. 7 the path taken by a light pulse, where the

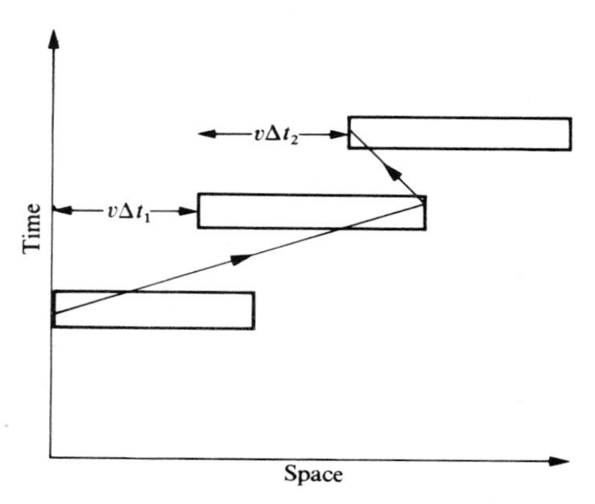

FIG. 7. Space–time diagram of the path taken by a pulse in the light pulse clock whilst travelling from one end to the other and back.

horizontal axis denotes distance and the vertical axis denotes time. Such a space–time diagram is a useful way of describing the motion in a compact form, and we will resort to it frequently.

Let us suppose that a stationary observer measures the length of the moving clock, and finds it to be l. This is achieved by the usual process of finding the distance between the two ends of the clock at a given time, as measured in the stationary frame of reference. We let Δt_1 be the time taken for the light pulse to reach the farther end of the clock, and Δt_2 to be the time taken for it to return to its original end, both measured by stationary clocks.

In the time Δt_1 the farther end of the clock has travelled a distance $v\Delta t_1$, so the distance travelled by the light pulse in the time Δt_1 is $(l+v\Delta t_1)$ (see Fig. 7). Since the light is travelling with velocity c then

$$c\Delta t_1 = l+v\Delta t_1. \tag{2.11}$$

For the return trip of the light pulse the original end of the clock is approaching the pulse, so the distance travelled by the pulse is $(l-v\Delta t_2)$. Since the light is still travelling with speed c, and it take the time Δt^2 to travel $(l-v\Delta t_2)$, we have

$$c\Delta t_2 = l - v\Delta t_2. \tag{2.12}$$

From (2.11) and (2.12) we obtain

$$\Delta t_1 = \frac{l}{c-v}, \qquad \Delta t_2 = \frac{l}{c+v},$$

so that the total time taken for the light pulse to travel to the farther mirror and back will be

$$\Delta t = \Delta t_1 + \Delta t_2 = \frac{2lc}{c^2 - v^2}. \tag{2.13}$$

We now use the result of §2.2 on the relation between the times in the two frames of reference. If Δt_0 is the time taken by the light pulse on its journey to the farther mirror and back, as measured by an observer travelling along with the clock, then by (2.4)

$$\Delta t = \Delta t_0 \left(1 - \frac{v^2}{c^2}\right)^{-\frac{1}{2}}. \tag{2.14}$$

But since the light pulse also travels with speed c as far as an observer moving with it is concerned,

$$\Delta t_0 = \frac{2l_0}{c}. \tag{2.15}$$

If we combine (2.13), (2.14), and (2.15) we have

$$\Delta t = 2l_0(c^2 - v^2)^{-\frac{1}{2}} = \frac{2lc}{c^2 - v^2},$$

so that

$$l = l_0 \left(1 - \frac{v^2}{c^2}\right)^{\frac{1}{2}}. \tag{2.16}$$

We obtain, from (2.16), a reduction in length of the moving clock, as viewed by a stationary observer, compared to its length when measured by someone travelling with the clock. This is the phenomenon of space contraction and can be expressed as 'moving rods get shorter'. If we take account of the other dimensions of an object we only expect contraction along the direction of motion. Thus moving objects should get flattened; a spherical particle will appear as ellipsoidal when viewed in relative motion. We should realise that the contraction is not usually seen by the eye. If account is taken of the time light takes to travel to a stationary observer's eye from different parts of a moving object it will be realised that the object will appear distorted.

If the object is very distant it appears to the eye as if rotated but without any contraction as such.

As in the case of time dilation the factor $(1-v^2/c^2)^{\frac{1}{2}}$ is very close to unity except at speeds very near that of light. It is only when $v \approx c$ that length contraction becomes appreciable. This is the case in respect of decaying muons described in the previous section; for them the contraction factor was $\frac{1}{9}$, and a spherical muon should appear like a nearly flat pancake to a stationary observer. Electrons at the Stanford two-mile-long linear accelerator have a value of about 2×10^{-5} for the contraction factor, so that they would look like very thin discs. There are also additional distortions which arise when looking at objects moving very fast.

We have already seen a validation of the phenomenon of space contraction in the Michelson–Morley experiment discussed in Chapter 1. A shortening of the arms of the interferometer in the direction of motion of the apparatus through the ether by an amount equal to (2.16) gave a perfect explanation of the null result.

This space contraction does not explain the null result of the further experiment with modified apparatus having arms of different length. We raised earlier the conjecture that time as well as space distortion is needed to comprehend this. We have now accepted the phenomenon of time dilation as well as space contraction. But these appear rather separate from each other. Our next task is to combine together these changes so that situations in which both occur can be discussed compactly. This will lead us to the Lorentz transformation of coordinates, and so to the heart of special relativity; our guiding principle will be the constancy of the speed of light. If we can achieve such a theory then there will be no need to explain the null result of the extended Michelson–Morley experiment, since that automatically follows from the constant speed of light.

Further reading for Chapter 2

On the time-dilation experiment with muons, see D. H. Frisch and J. H. Smith, *Am. J. Phys.* **31**, 342–55 (1963).

On time dilation for clocks travelling on jet planes, J. C. Hafele and R. E. Keating, Around the world atomic clocks: predicted relativistic time gains, and, Around the world atomic clocks: observed time gains, *Science* **177**, 166–70 (1972).

On the difficulties of accepting time dilation (for incorrect reasons), and an interesting case study in the psychology of scientists, see H. B. Dingle, *Science at the crossroads*, Martin Brian and O'Keefe, London (1972).

On observing the distortion of moving objects to the eye, see J. Terrell, *Phys. Rev.* **116**, 1041–5 (1959); V. F. Weisskopf, *Phys. Today*, **13**, 24–27 (1960), and G. D. Scott and M. R. Viner, *Am. J. Phys.* **33**, 534 (1965).

3. The Lorentz transformation

3.1. Observers and events in space–time

AN event is something that happens at a given point in space at a given time. It may be a measurement of some quantity, it may be the interaction of two or more sub-nuclear particles, or it may arise from the persistence of an object at certain point or moving with a certain speed. It is convenient to picture an event as a point in a space–time diagram. We live in a space of three dimensions, but it is necessary to delete two of these dimensions in order to be able to draw space–time diagrams conveniently. The resulting pictures are of help in visualizing what may be happening in the full three-dimensional world, though care must be taken in transposing such understanding.

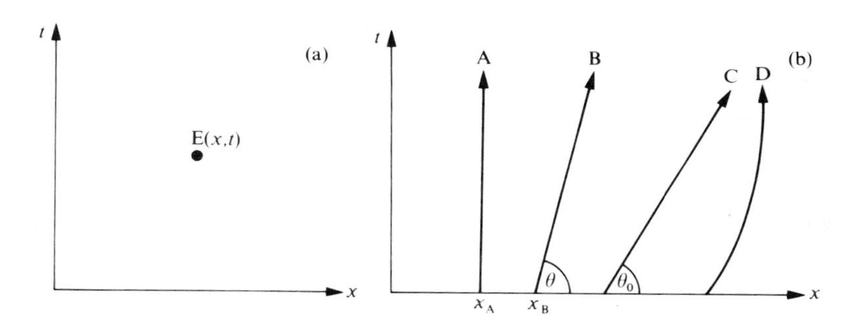

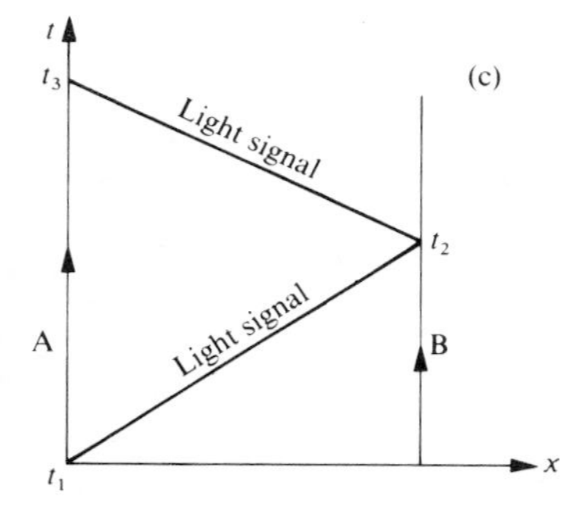

FIG. 8. Space–time diagrams (a) of an event at point x at time t; (b) of the world line A of a stationary particle at x_A; the world line B of a particle moving with velocity $\cot\theta$ to the right; the world line C of a pulse of light; the world line D of a particle initially moving close to the speed of light and then slowing down. (c) Synchronization of a distant clock at B by light signals.

A one-dimensional space-time diagram is shown in Fig. 8(a). The event E at the point x in space at time t is thus represented by the point with co-ordinates (x, t) in this diagram. The infinite set of events corresponding to an object persisting at the point x_A in space is represented by the vertical line A in the space–time diagram of Fig. 8(b), and the set of events corresponding to the object moving with velocity v and passing through the point x_B in space at time $t = 0$ is described by the line B of Fig. 8(b), with $v = \cot\theta$. The line C at the angle $\theta_0 = \cot^{-1} c$ to the x-axis corresponds to a ray of light travelling to the right, while D represents a particle moving initially at very close to the speed of light and then slowing down to rest. These curves A, B, C, D are called the world lines of the corresponding objects.

The values of x and t of an event are assumed to be measurable with arbitrary accuracy by means of a set of synchronized clocks and rigid rods with which the particular frame of reference is provided. Further, there are supposed to be an arbitrary number of observers to man such measuring apparatus so that the lines A, B, C, D shown in Fig. 8(b) or any other infinite sets of events can be plotted, again with arbitrary accuracy. The process of synchronization of distant clocks can be achieved by light signals sent from one of the clocks to another, as shown in Fig. 8(c). The time t_2 of the distant clock B is defined to be synchronized with the clock A if B reads the mean of the time t_1 and t_3 at which a light signal is sent from A and returns to A after being reflected from B;

$$t_2 = \tfrac{1}{2}(t_1 + t_3).$$

3.2. The relativity of simultaneity

Two events are said to be simultaneous if they occur at the same time. In the previous chapter it was found that time suffered dilation on motion: 'moving clocks go slow relative to stationary observers'. But the modification of time by motion could well destroy the absolute character of simultaneous events; they may only appear simultaneous when observed at a suitable speed.

To analyse this possibility let us first consider three stationary observers A, B, C an equal distance apart, with B in the middle (see Fig. 9(a)). At time $t = 0$ observer B flashes two light signals, one towards A, the other to C. The receipt of these signals by A and C constitute the events A_1 and C_1. The definition of synchronization of stationary clocks in the previous section allows us to conclude that A_1 and C_1 are simultaneous.

Let us now consider A, B, and C moving with the same constant speed v and let B repeat the exercise of sending out the two short pulses of light in opposite directions. The space–time diagram of the situation is now shown in Fig. 9(b); since the velocity of light is unchanged the space–time locus of the light pulses are unchanged from the situation of Fig. 9(a). It is clear that

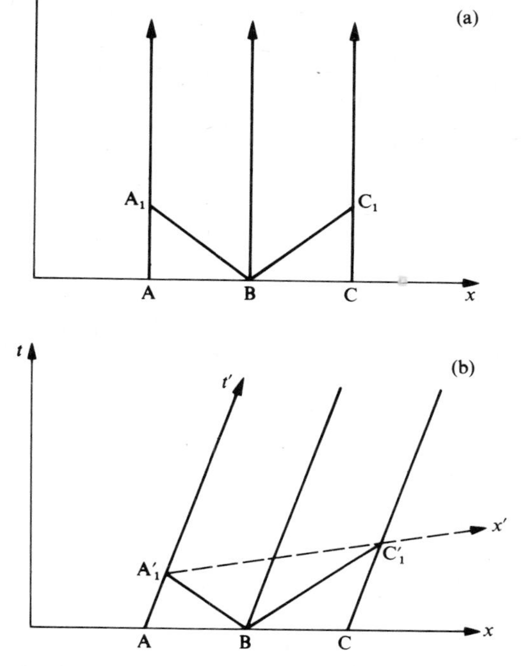

FIG. 9. (a) Light signals sent out from B to equidistant observers A and C are received simultaneously. (b) If the observers A and C are all in uniform motion the receipt of the signals from B by A and C no longer appears simultaneous.

the events A'_1, C'_1 of receipt of the pulses by A and C are no longer simultaneous, since they occur at different times. However, they certainly are regarded as simultaneous to the observers A, B, and C and to any observers moving with them at the same speed, since to them the situation is still as in Fig. 9(a).

This analysis makes clear the relative nature of simultaneity in a world in which light always travels at the same speed. What is more, it is necessary to introduce a new time measured along the coordinate axis t' and distances measured along x', as shown in Fig. 9(b), which correspond to the frame of reference moving with the observers A, B, C.

3.3. The Lorentz transformation

Let us consider in more detail the manner in which space and time values are changed by motion. We realized in the previous chapter that such modification occurs in the phenomena of space contraction and time dilation. However, we only considered situations where one or other of them occurred, but not both. In § 3.2 we discovered that if motion occurs both space and time have to be altered together. That was made especially clear by Fig. 9(b). For the time t' measured in the frame moving with A, B, and C must be parallel to the world lines of these three observers since they are at rest in the moving frame. Similarly the x'-coordinate axis must be parallel to $A'_1C'_1$, since that is the line of simultaneity in the moving frame.

We must now attempt to find an algebraic relation between the coordinates (x, t) of an event E in the stationary frame S and its coordinates (x', t') in the moving frame. To do that we see first that the relation between (x, t) and (x', t') must be a linear one, since rectilinear motion in one frame remains so when viewed from the other frame. It could not do so if x' or t' were non-linear functions of x and t. Therefore, there must exist constants a, b, d, e, f, and g (we do not use c since that is reserved for the velocity of light) so that for all corresponding sets of coordinates (x, t), (x', t') of the same event E in the two different frames

$$x' = ax - bt + f, \tag{3.1}$$

$$t' = dx + et + g. \tag{3.2}$$

If we require that the origins of the two coordinate frames S and S′ coincide then we must have in (3.1) and (3.2)

$$f = g = 0. \tag{3.3}$$

We now impose the condition that the origin 0′ of S′, the point $x' = 0$, travels with velocity v in S. We obtain from (3.1) that the coordinates (x, t) in S of 0′ must satisfy

$$ax - bt = 0,$$

so that 0′ moves with velocity $x/t = b/a$. Thus

$$b = av. \tag{3.4}$$

We must also require that the velocity in S′ of the origin 0 of S, the point $x = 0$, is $-v$. From (3.1) and (3.2) the motion of 0 in S′ is described by

$$x' = -bt, \qquad t' = et,$$

so 0 will have velocity $x'/t' = -b/e$ in S′. Therefore

$$b = ev. \tag{3.5}$$

We then require that a pulse of light travels with velocity c in both S and S′. In other words, the relation $x = ct$ is equivalent of $x' = ct'$. Using (3.1) and

(3.2) this becomes

$$ct' = act - bt, \tag{3.6}$$

$$t' = dct + et. \tag{3.7}$$

These two equations must be valid for all corresponding values of t and t'; taking their ratio we obtain

$$c = \frac{ac - b}{cd + e}$$

or

$$c^2 d = -b. \tag{3.8}$$

If we now combine (3.1), (3.2), (3.3), (3.4), (3.5), and (3.8), we obtain

$$x' = a(x - vt), \tag{3.9}$$

$$t' = a\left(t - \frac{vx}{c^2}\right). \tag{3.10}$$

Finally we have to obtain the constant a as a function of v and c. We can do that by changing the direction of the coordinate axes x and x' in S and S′, so that now S moves with a velocity v with respect to S′ (see Fig. 10). The roles of the two frames are now interchanged, so the relations (3.9) and (3.10) should still hold, but with x replaced by $-x'$, x' replaced by $-x$, and t and t' interchanged. In other words,

$$-x = a(-x' - vt'), \tag{3.11}$$

$$t = a\left(t' + \frac{vx'}{c^2}\right). \tag{3.12}$$

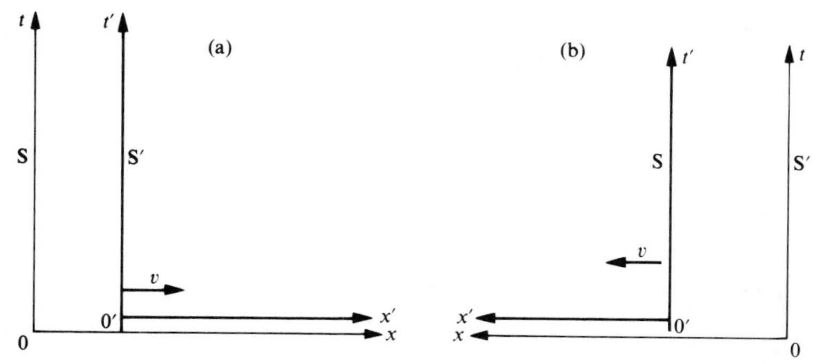

FIG. 10. (a) Relation between the frames S and S′ used in considering the Lorentz transformation. (b) The frames S and S′ after interchange of x and $-x'$.

Since (3.9)–(3.12) must hold simultaneously for all pairs of (x, t), (x', t') describing the same event E, we find that substituting for x and t from (3.11) and (3.12) into (3.9)

$$x' = a\left\{a(x'+vt') - va\left(t' + \frac{vx'}{c^2}\right)\right\} = a^2x'\left(1 - \frac{v^2}{c^2}\right). \tag{3.13}$$

Thus

$$a = \left(\frac{1-v^2}{c^2}\right)^{-\frac{1}{2}}, \tag{3.14}$$

where we choose the positive square root in (3.14) so that when $v = 0$ then $x' = x$ and $t' = t$. Thus we finally obtain the explicit relation

$$x' = \left(1 - \frac{v^2}{c^2}\right)^{-\frac{1}{2}}(x - vt), \tag{3.15}$$

$$t' = \left(1 - \frac{v^2}{c^2}\right)^{-\frac{1}{2}}\left(t - \frac{vx}{c^2}\right). \tag{3.16}$$

The transformation from (x, t) to (x', t') given by (3.15) and (3.16) is called a Lorentz transformation. It is this which gives us the final answer to the question as to how space and time are both altered by motion; the Lorentz transformation is the basis of special relativity.

It is usual to denote the factor $(1 - v^2/c^2)^{-\frac{1}{2}}$ py $\gamma(v)$, and we will do that here. Then the Lorentz transformation (3.15), (3.16) is written

$$x' = \gamma(v)(x - vt), \tag{3.17}$$

$$t' = \gamma(v)\left(t - \frac{vx}{c^2}\right), \tag{3.18}$$

and the inverse transformations (3.11), (3.12) become

$$x = \gamma(v)(x' + vt'), \tag{3.19}$$

$$t = \gamma(v)\left(t' + \frac{vx'}{c^2}\right). \tag{3.20}$$

We will use the forms (3.17)–(3.20) as the Lorentz transformation equations from now on.

3.4. Derivation of time dilation and space contraction

The rules (3.15) and (3.16) for the transformation of coordinates has been based upon the requirement that the velocity of light is a constant c, independent of how it is observed. This condition was used in the previous chapter to derive time dilation. It should be possible to derive this slowing of time from (3.15) and (3.16). We proceed now to do this.

A clock moving with velocity v with respect to a fixed frame of reference S can be considered as being at rest in a frame S′ moving with velocity v with respect to S. It will travel along the line A in the space–time diagram of Fig. 11, where x and t are the space and time coordinates of S and the angle between A and the x-axis will be $\cot^{-1} v$.

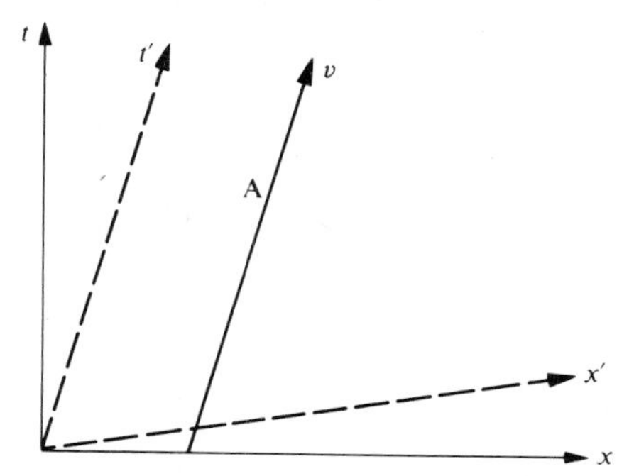

FIG. 11. The world line A of a moving clock. The time axis in a frame at rest with respect to the clock is parallel to A.

In the frame S′ the clock A has a constant value of the space coordinate x'. Thus the time-axis t' of S′ must be parallel to A, as shown in Fig. 11. The x'-axis will be at an angle θ to the x-axis, which may be determined from (3.10) by putting $t' = 0$ (the equation of the x'-axis). The x'-axis is thus $t = vx/c^2$, so $\theta = \tan^{-1}(v/c^2)$.

It may be helpful to digress briefly on this matter of oblique axes. The coordinates (x, t) of an event E as measured in the frame S are represented by dropping perpendiculars to the two axes, as shown in Fig. 12. This is

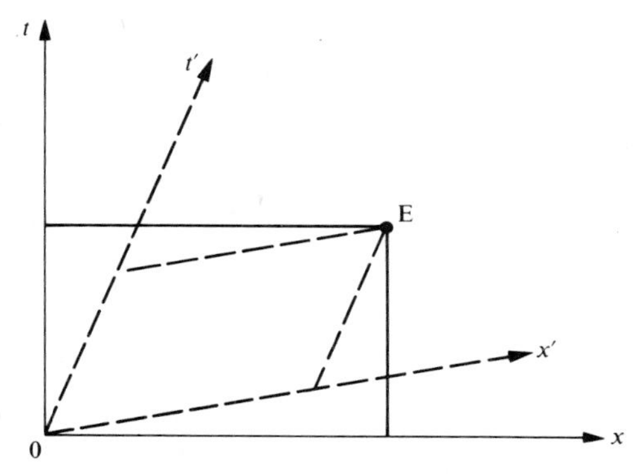

FIG. 12. Coordinates of an event E in rectangular and oblique axes.

correct when we represent the two coordinate axes as being at right angles. The coordinates (x', t') of E in the frame S′ moving with velocity v with respect to S cannot be obtained in a similar way since we have seen that the x'- and t'-coordinate axes are oblique. The (x', t')-coordinates of E with respect to oblique axes are determined by the lengths of the edges of the parallelogram whose sides are parallel to the x'- and t'-coordinate axes and with one vertex at the origin 0, the other at the event E itself (see Fig. 12). Such a prescription evidently reduces to the usual one of dropping perpendiculars when the axes are orthogonal. The use of oblique axes is essential in relativity if we wish to represent the coordinates of the same event as measured in different frames. However, care must be taken to avoid mistakes in using such axes. For example, one cannot think of distances between two points as given by that obtained by direct measurement. With such provisos, oblique axes are a great aid in picturing geometrically the nature of the Lorentz transformation, and we will use it constantly.

Let us now return to the problem of the time dilation. Suppose the clock A is at the point with x'-coordinate x'_A, as measured in the frame S′. Let t'_1, t'_2 be two times measured on it in S′, with corresponding times as measured from the stationary frames S being t_1, t_2. Then by the inverse Lorentz transformation equations (3.19) and (3.20),

$$t_1 = \gamma(v)\left(t'_1 + \frac{vx'_A}{c^2}\right), \tag{3.21}$$

$$t_2 = \gamma(v)\left(t'_2 + \frac{vx'_A}{c^2}\right). \tag{3.22}$$

Subtracting (3.21) from (3.22) we obtain

$$(t_2 - t_1) = \gamma(v)(t'_2 - t'_1). \tag{3.23}$$

Eqn (3.23) is identical to the time-dilation formula (2.4) of Chapter 2, specifying the increased time, as measured in S, between two events in S′. If the two events in S′ are the consecutive beats of a clock, stationary in S′, then this clock will appear to go slow as seen in S; the time-dilation factor is $\gamma(v)$, as we saw earlier.

By the same argument used for time dilation we expect that the space-contraction phenomenon discussed in Chapter 2 should also result from the Lorentz transformation law (3.17)–(3.20). We have first to specify how lengths of moving objects are to be measured. If we have a rod at rest we will define its length, when measured in a stationary frame S, as the distance between its ends when observed at the same time in S. Similarly, its length in a frame S′ moving with respect to the rod with velocity v will be the distance between its ends when observed at the same time in S′.

Let the ends of the rod have the x-coordinates x_1, x_2 in the fixed frame S. To find the length of the rod in the moving frame S′ we must measure the positions of its ends at the same time t'_0 in S′. We denote the x'-coordinates of the ends as x'_1 and x'_2 when these measurements are made; the events $E_1 = (x'_1, t'_0)$ and $E_2 = (x'_2, t'_0)$ are shown in Fig. 13, which is drawn so that

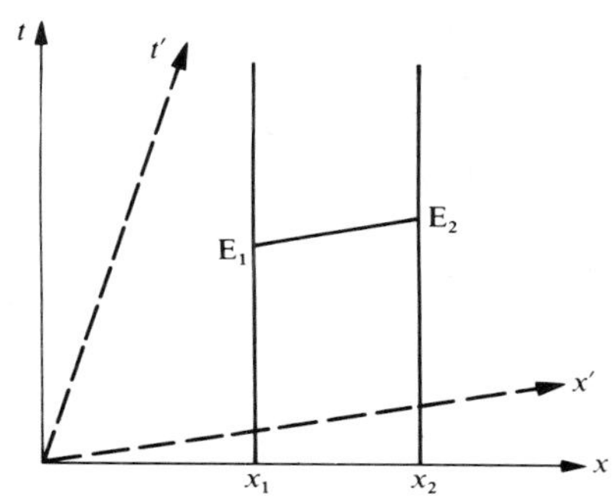

FIG. 13. Measuring the length of a rod while in motion. The events E_1, E_2 are simultaneous in the moving frame S′, so allowing values of the x'-coordinates of the ends of the rod to be related to x_1 and x_2, the coordinates in the stationary frame.

the line E_1E_2 is parallel to the x'-axis and has a constant value of t'. By the Lorentz transformation (3.19) and (3.20)

$$x_1 = \gamma(v)(x'_1 + vt'_0), \tag{3.24}$$

$$x_2 = \gamma(v)(x'_2 + vt'_0). \tag{3.25}$$

Substracting (3.24) from (3.25) gives

$$(x_2 - x_1) = \gamma(v)(x'_2 - x'_1). \tag{3.26}$$

Since $(x_2 - x_1)$ is the length l of the rod in S and $(x'_2 - x'_1)$ is its length l' in the moving frame S′ we have

$$l' = \left(1 - \frac{v^2}{c^2}\right)^{\frac{1}{2}} l,$$

which is the space contraction formula (2.16) of Chapter 2.

3.5. Space–time and causality

We have seen how to relate the coordinates of an event as observed from two relatively moving frames of reference by means of the Lorentz transformation. We have given this relationship a geometrical interpretation as a change of axes, as was clear in Fig. 11 (p. 29) and the discussion on it in the previous section. It is possible to implement this geometrical approach even further, and we will now do so because it helps to obtain a very important appreciation of the causal structure of space and time.

A very valuable quantity for any event E is the value

$$c^2t^2 - x^2. \tag{3.27}$$

This expression, quadratic in the coordinates x and t of the event E, would seem to depend heavily on the frame S used to measure these coordinates. In fact this is not so, and (3.27) is completely independent of the frame S (except for the choice of origin). To see this let us calculate its value in terms of any other coordinates (x', t') of E measured in another frame S′ moving with velocity v with respect to S and with the same origin. We use the Lorentz transformation (3.19) and (3.20) to express (3.27) in terms of x' and t' for E,

$$\begin{aligned} c^2t^2 - x^2 &= c^2\gamma^2(v)\left(t' + \frac{vx'}{c^2}\right)^2 - \gamma^2(v)(x' + vt')^2 \\ &= \gamma^2(v)\left\{(c^2 - v^2)t'^2 + 2vt'x' - 2vt'x' + x'^2\left(\frac{v^2}{c^2} - 1\right)\right\} \\ &= c^2t'^2 - x'^2, \end{aligned} \tag{3.28}$$

where we use the value $\gamma(v) = (1 - v^2/c^2)^{-\frac{1}{2}}$. We have thus proved in (3.28) that $(c^2t^2 - x^2)$ is an invariant, being independent of the choice of reference frame S, except for the choice of origin.

The label $(c^2t^2 - x^2)$ may be attached to an event in a coordinate-independent fashion. We may use it to classify the possible events in space–time. In particular, we may divide events into three sorts, being those for which the invariant label is positive, zero, or negative. Events for which the label is positive are called timelike with respect to the origin, since they are of the class containing those with $x = 0$ and $t \neq 0$, which are just changes in time of a clock at the origin of space. Those for which the invariant is negative are called spacelike with respect to the origin since that includes events with $t = 0$, $x \neq 0$, which are just spatially separated but simultaneous with an event at the origin of space–time. The final class of events with the invariant label (3.27) being zero are called lightlike with respect to the origin because a ray of light can pass to or from the origin of space–time to them. This division of space–time into three regions is shown in Fig. 14, and is expressed as

(1) $c^2t^2 - x^2 > 0$: timelike with respect to the origin;
(2) $c^2t^2 - x^2 < 0$: spacelike with respect to the origin;
(3) $c^2t^2 - x^2 = 0$: lightlike with respect to the origin.

Such a classification of events can also be considered from the point of view of causality. We already remarked in Chapter 2 that it did not seem possible to have a frame moving with respect to another one at a speed greater than

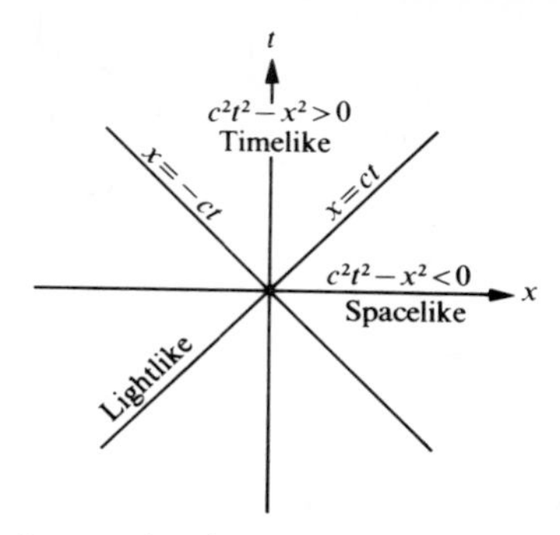

FIG. 14. The division of space–time into three different regions of events which are timelike, lightlike, or spacelike with respect to the origin.

that of light. This feature follows in more generality from the Lorentz transformation (3.17)–(3.20) since if $v > c$ then $\gamma(v)$ would be imaginary and one or other of the two frames would have imaginary spaces and times. We expect that this limitation $v \leqslant c$ will apply to any material particle which, at some time or another, has been travelling at a speed satisfying the condition. For then we should be able to attach a coordinate frame to it, and such a frame would suffer the same fate of turning imaginary if the particle could ever be accelerated to travel faster than light.

We conclude that information can only travel at speeds less than or equal to c. The existence of faster-than-light particles would modify this, but since these have not been found, and there are even very strong theoretical reasons against their existence (which will be discussed in Chapter 8), we will elevate this result to a principle, the *principle of causality*:

information cannot travel faster than light.

It results from this principle that an event E can be influenced by an event at the origin O—it is causally connected to O—if and only if E is timelike or lightlike with respect to O. For only then could a particle or ray of light travel from O to E; if E were spacelike with respect to O only a faster-than-light particle could reach E from O. Thus space–time is divided causally into two parts, those events E which are causally connected to O and those which cannot be.

The set of events E which are timelike or lightlike with respect to O is called the light cone of O. In a two-dimensional space–time this set is not a cone at all, but on rotation about the t-axis one obtains such a cone in a three-dimensional space–time of two space and one time dimensions, as in Fig. 15. A slightly more complicated volume is obtained in three space and one time dimensions, but it is still a cone (we will return to three space dimensions in

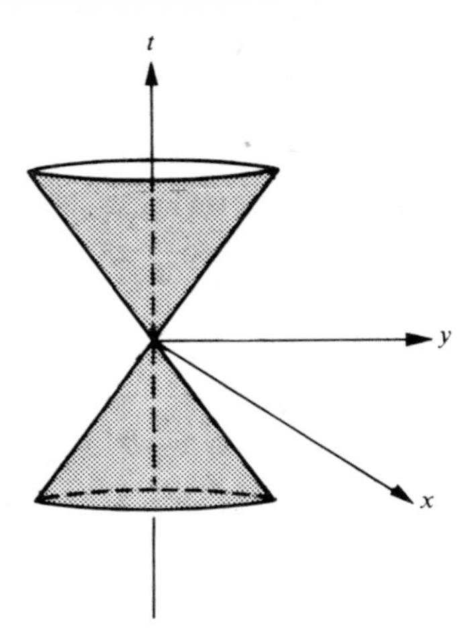

FIG. 15. The future and past light cones with vertex at the origin, in a two-dimensional space.

Chapter 4). The surface of the light cone is composed of events which are lightlike with respect to O, and its interior of events which are timelike to O. The set of timelike and lightlike events with a positive time is called the future light cone, and that set of events with negative time the past light cone. Thus

$$[(x, t): c^2t^2 - x^2 \geqslant 0, t > 0] = \text{future light cone},$$

$$[(x, t): c^2t^2 - x^2 \geqslant 0, t < 0] = \text{past light cone}.$$

We are using the notation $[(x, t): P(x, t)]$ for the set of events with coordinates (x, t) for which the property $P(x, t)$ is true. Finally, O is called the vertex of the light cone.

At any event E in space–time we may construct the light cone with E as vertex. By this we have divided the set of events into those which are timelike, lightlike, or spacelike with respect to E. As before, the set of events causally connected to E are contained in the light cone with vertex at E. It is clear that those events which can influence E are contained in the past light cone at E, and those which can be influenced by E are in the future light cone at E. In this manner the complete causal structure of space–time is made explicit.

Finally it is useful to note that the different coordinates of an event E, as measured in different frames, are constrained by (3.28). It is sometimes possible to use the invariant (3.27) alone to solve problems, without going to the full Lorentz transformation. When we turn to analyse Lorentz transformations in three dimensions in Chapter 7 we will see that the invariant provides a very large amount of information for analysing the theoretical framework of the material world.

Further reading for Chapter 3

A simple but brilliant account of Lorentz transformations is given by Sir Hermann Bondi in *Relativity and common sense*, Science Study Series No. 31, Heinemann, London (1965). There is also an excellent account, as should be expected, in Chapters 1 and 2 of Albert Einstein's *The meaning of relativity*, Methuen Science Paperbacks, London (1967).

4. Relativistic kinematics

4.1. Relative velocities

THE Lorentz transformation (3.17)–(3.20) modifies the space and time coordinates of events measured from one coordinate frame as compared to those measured in one moving relatively to the first. The velocity of an object, being the distance it has travelled per unit time, is also to be expected to be modified. The velocity transformation law cannot be the Galilean one (1.3) since that would be in contradiction with the constancy of the speed of light in the case where the moving object investigated is light itself. We will begin our discussion of relativistic kinematics by determining how the velocity of a particle is modified when it is measured from a moving frame.

The velocity u of a particle, with space–time coordinates (x, t), as measured in a frame S, is

$$u = \frac{\mathrm{d}x}{\mathrm{d}t}. \tag{4.1}$$

Measured in a frame S′ moving with velocity v with respect to S the particle will have a velocity we denote by u' and space–time coordinates (x', t'), with

$$u' = \frac{\mathrm{d}x'}{\mathrm{d}t'}. \tag{4.2}$$

Differentiating (3.19) and (3.20) with respect to t' and x' respectively, we obtain

$$\frac{\mathrm{d}x}{\mathrm{d}t'} = \gamma(v)\left(\frac{\mathrm{d}x'}{\mathrm{d}t'} + v\right), \tag{4.3}$$

$$\frac{\mathrm{d}t}{\mathrm{d}x'} = \gamma(v)\left(\frac{\mathrm{d}t'}{\mathrm{d}x'} + \frac{v}{c^2}\right). \tag{4.4}$$

Now

$$u = \frac{\mathrm{d}x}{\mathrm{d}t'} \frac{(\mathrm{d}x'/\mathrm{d}t)}{u'}, \tag{4.5}$$

so substituting (4.3) and (4.4) in (4.5) we have

$$u = \gamma(v)(u'+v)/u'\gamma(v)\left(\frac{1}{u'} + \frac{v}{c^2}\right) = (u'+v)\bigg/\left(\frac{1+u'v}{c^2}\right). \tag{4.6}$$

Eqn (4.6) is the required transformation formula determining u when u' is known. Solving for u' we obtain the inverse transformation

$$u' = \frac{(u-v)}{1-(uv/c^2)}. \tag{4.7}$$

One of the important properties of (4.6) is that whatever values of u' and v are chosen, with absolute value not greater than c, then the absolute value of u is also not greater than c. For if we introduce parameters a, a', so that

$$v = ac, u' = a'c, \qquad |a| \leqslant 1, |a'| \leqslant 1 \tag{4.8}$$

then

$$\frac{u}{c} = \frac{a+a'}{1+aa'}, \qquad 1-\frac{u}{c} = \frac{(1-a)(1-a')}{1+aa'}. \tag{4.9}$$

But from (4.8)

$$0 \leqslant (1-a), \qquad 0 \leqslant (1-a'), (1+aa') \geqslant 0,$$

so the minimum value of $(1-a)(1-a')/(1+aa')$ is zero. The positions of the maximum value of the function are obtained by equating to zero its partial derivatives with respect to a and a'; the partial derivatives are

$$\frac{(a'-1)(a'+1)}{(1+aa')^2} \quad \text{and} \quad \frac{a^2-1}{(1+aa')^2}. \tag{4.10}$$

These are zero at $a = \pm 1$ and $a' = \pm 1$; the two values of the function $(1-a)(1-a')/(1+aa')$ at these four points are 0 and 2, which are clearly minima and maxima respectively. Thus

$$0 \leqslant 1-(u/c) \leqslant 2,$$

so

$$-1 \leqslant u/c \leqslant 1 \quad \text{or} \quad |u| \leqslant c.$$

Thus it is never possible to make a particle travel faster than light by projecting it at a suitably high speed from a rapidly moving object. For example, if both v and u' are taken to be $0{\cdot}8c$ then by (4.6) $u = 0{\cdot}975c$, though $v+u' = 1{\cdot}6c$. The difference between the relativistic transformation law (4.6) and the Galilean equation (1.3) is quite clear.

When the transformation laws are applied to the case of light we still obtain agreement with the fundamental condition of the constancy of light. Thus if $u = c$ in (4.7) then

$$u' = \frac{c-v}{1-v/c} = c,$$

while if $u' = c$ in (4.6) then

$$u = \frac{c+v}{1+v/c} = c,$$

as required.

What is also satisfactory about the velocity transformation laws (4.6) and (4.7) is that they reduce to the corresponding Galilean ones when the relative velocity v of the two frames S and S′, as well as the velocities u or u' respectively are much less than c. For then we can neglect the denominators in each of these equations, and obtain the Galilean transformation laws immediately.

4.2. The Doppler effect

The Doppler effect is the change of frequency of radiation from its value seen at rest to that seen by an observer with respect to whom the source of the radiation is in motion. This is a phenomenon first noticed by Christian Doppler in 1842 and often experienced, in the age of the motor car, as a sudden drop in pitch of the noise of a car or siren as it passes. The effect can be investigated at either the non-relativistic or relativistic level. We will commence by studying the former situation before turning to the latter.

Consider some form of vibration, such as a sound wave, emitted by a source moving with velocity u_1, and being absorbed by a receiver travelling with speed u_2. We assume that the vibration travels with velocity v and has a frequency ν and period τ, so $\nu = \tau^{-1}$. Owing to the motion of the source the distance between the crest of the waves will be modified; the source will have moved a distance $u_1\tau$ between the emission of one crest and the next. Thus the effective wavelength λ of the radiation will be $(v-u_1)\tau$, taking the direction in which the source is moving to be the same as that of the radiation. The speed of the pulses relative to the receiver will be $(v-u_2)$. The period τ' of the radiation as observed by the receiver will thus be the time taken to travel a distance λ with velocity $(v-u_2)$, and so will be $\lambda/(v-u_2)$. Thus the observed frequency ν' of the vibration will be

$$\nu' = \frac{v-u_2}{\lambda} = \frac{v-u_2}{(v-u_1)\tau} = \frac{\nu(v-u_2)}{v-u_1}. \tag{4.11}$$

The modification of frequency specified by (4.11) is the Doppler effect. In particular, when the source is stationary but the observer is moving the modified frequency is $\nu(1-u_2/v)$, whilst if only the source is moving the new frequency becomes $\nu(1-u_1/v)^{-1}$. Thus if the source is travelling towards the observer both u_1 and v must have the same sign so that there is an increase of frequency; if the source is receding then u_1 and v have opposite signs and the frequency is reduced.

Let us now turn to an analysis of the Doppler effect for high-speed sources or observers, when the Lorentz transformation (3.17)–(3.20) is necessary to relate space and time coordinates of the associated relatively moving frames.

We restrict our analysis to the effects of motion on electromagnetic radiation, travelling with velocity c. Suppose there is a source of brief pulses of light placed at the origin and emitting pulses every τ s. Then the path of a particular light ray, which we consider as the first of a sequence of $(n+1)$ pulses, will be

$$x = ct.$$

The path of the $(n+1)$th pulse will be obtained by delaying time by $n\tau$, and so will be

$$x = c(t-n\tau).$$

Let us consider an observer moving with velocity v with respect to the source and starting at $t = 0$ a distance x_0 from the origin. The path of the observer will be described by

$$x = x_0 + vt.$$

Then the first light pulse is received by the observer at the space–time co-ordinates (x_1, t_1) with

$$x_1 = ct_1 = x_0 + vt_1, \tag{4.12}$$

and the $(n+1)$th at (x_2, t_2) with

$$x_2 = c(t_2 - n\tau) = x_0 + vt_2. \tag{4.13}$$

From (4.12) and (4.13) we obtain

$$(t_2 - t_1) = \frac{cn\tau}{c-v}, \qquad (x_2 - x_1) = \frac{cnv\tau}{c-v}.$$

The time between the first and $(n+1)$th pulses being received by the observer, as measured by him, are thus given from the Lorentz transformation (3.20) as

$$(t'_2 - t'_1) = \gamma(v)\left\{(t_2 - t_1) - \frac{v(x_2 - x_1)}{c^2}\right\} = \frac{\gamma(v)cn\tau(1 - v^2/c^2)}{c - v}.$$

Then the period of the pulses that the moving observer measures by his own clocks will be

$$\tau' = \frac{(t'_2 - t'_1)}{n} = \gamma(v)\frac{c\tau(1 - v^2/c^2)}{c - v} = \left(1 + \frac{v}{c}\right)^{\frac{1}{2}} \tau \left(1 - \frac{v}{c}\right)^{-\frac{1}{2}},$$

and the corresponding frequency will be

$$\nu' = \{(1-\beta)(1+\beta)^{-1}\}^{\frac{1}{2}}\nu, \tag{4.14}$$

where $\beta = v/c$. Eqn (4.14) is the relativistic version of the Doppler-effect formula (4.11).

The difference between the relativistic and non-relativistic formulae becomes important at high speeds, such as those obtained by galaxies in their common recession. Thus galaxies have been observed which have values of v/c, as deduced from (4.14), as large as 0·9. For them the factor ν'/ν, equal to $\{(1-\beta)/(1+\beta)\}^{\frac{1}{2}}$, is 0·23, whilst the non-relativistic formula (4.11) with $u_2 = 0$, $u_1/v = 0{\cdot}9$ would give a value of ν'/ν of $(1+\beta)^{-1}$ equal to 0·49. It is only when $\beta \ll 1$ that the non-relativistic and relativistic formulae give nearly equal Doppler shifts. Care must be taken, however, in applying (4.14) to very distant galaxies which have such high speeds since a model of the Universe is necessary to account for additional gravitational effects.

4.3. Light in a moving medium

When light passes through a medium of refractive index n its velocity is reduced to the value c/n. This is consistent with the constancy of the velocity of light from which we have derived the Lorentz transformation, since light is only required to have a constant speed in a vacuum. When a transparent medium is present, light will have a velocity through it which will depend on relative velocities. We can calculate this velocity by means of the rule (4.6) for the transformation of velocities.

We suppose that the medium has velocity v along the direction of the light beam with respect to a stationary frame of reference. Relative to the medium light has velocity c/n, so that is the value for u' to be used on the right-hand side of (4.6). Therefore we obtain from that equation the velocity u of light in the moving medium, as measured by a stationary observer;

$$u = \left(\frac{c}{n}+v\right)\bigg/\left(1+\frac{v}{cn}\right). \tag{4.15}$$

We may expand the exact expression (4.15) in powers of v/c if $v/c \ll 1$, obtaining

$$\begin{aligned} u &= \frac{c}{n}\left(1+\frac{nv}{c}\right)\left(1+\frac{v}{cn}\right)^{-1} \\ &\approx \frac{c}{n}\left(1+\frac{nv}{c}\right)\left(1-\frac{v}{nc}\right) \\ &\approx \frac{c}{n}+\left(1-\frac{1}{n^2}\right)v. \end{aligned} \tag{4.16}$$

We may interpret the right-hand side of (4.16) as the sum of the velocity c/n of light in the medium plus some fraction f of the velocity v of the medium with respect to the stationary observer. This latter term indicates that there is some apparent dragging of the light by the medium; the factor f, equal to $\{1-(1/n^2)\}$, is called the dragging coefficient.

The dragging coefficient for water was measured by Fizeau in 1851 and found to be in good agreement with the value calculated from the refractive index; later experiments confirmed this. The drag hypothesis was originally explained by supposing that inside a moving transparent material light is carried partly by the material and partly by the ether penetrating it. Since the ether remains at rest (as found by stellar aberration) the light behaves as if only a proportion of the velocity of the material were added to that of light. This explanation of the dragging coefficient was destroyed by the Michelson–Morley experiment. It is satisfactory to see it being understood very simply by means of the laws of transformation of velocities based on the Lorentz transformation.

4.4. Three-dimensional Lorentz transformations

We have so far neglected any possible transformation of lengths measured at right angles to the direction of motion. In three dimensions we need to add to the Lorentz transformations (3.17)–(3.20) the relation between the y- and z-coordinates of an event as measured in the two relatively moving frames S, S′.

We expect that the relation between the three space coordinates (x, y, z) and the time coordinate t of an event E as measured in S will be linearly related to its coordinates (x', y', z', t') measured in the frame S′. For only by such a relation will rectilinear motion as observed in S remain rectilinear as seen from S′. We must proceed to find the coefficients of the linear transformation between these two sets of space–time coordinates (x, y, z, t) and (x', y', z', t'), under the condition that a pulse of light has always the same speed c.

We can see immediately that the speed of light in a direction at right angles to the relative motion of S and S′ will not be affected by such motion. Thus the constancy of the speed of light in motion along the y- and z-axes will be preserved if there is no distortion of these transverse coordinates,

$$y' = y, \qquad z' = z. \tag{4.17}$$

Indeed, we see that a pulse of light originally at the origin will expand as a spherical surface of radius r after time t with $r = ct$. This will be the surface

$$r^2 - c^2t^2 = x^2 + y^2 + z^2 - c^2t^2 = 0. \tag{4.18}$$

If we use the transformations (3.17), (3.18), and (4.17) this becomes

$$r'^2 - c^2t'^2 = x'^2 + y'^2 + z'^2 - c^2t'^2 = 0, \tag{4.19}$$

which is the equation of a spherical surface of radius ct', as observed in S′. This shows that light also travels with velocity c in S′ as well as in S.

We thus have the complete Lorentz transformation law in three dimensions,

$$\begin{aligned} x' &= \gamma(v)(x-vt) \\ y' &= y \\ z' &= z \\ t' &= \gamma(v)\left(t-\frac{vx}{c^2}\right). \end{aligned} \tag{4.20}$$

We have proved that the quantity $(x^2+y^2+z^2-c^2t^2)$ is unchanged by the Lorentz transformation (4.20); it is an invariant. This extends the invariant $(x^2-c^2t^2)$ which we discussed in § 3.5. We have used this implicitly already in our analysis of causality in a three-dimensional world; we have now justified those earlier remarks.

So far we have considered the very special case in which the only relative motion between two frames of reference is along their x-axes, which are parallel. Let us now try to extend this discussion to more general situations in which the relative motion is in a general direction, and the two sets of axes in space are not parallel. We will consider here only the former of these two cases, and defer discussion of the latter to a more appropriate place.

In order to extend the Lorentz transformation to arbitrary relative velocity which we denote by the three-vector $\boldsymbol{v}$, let us write (4.20) so that only three-vectors and time enter. The vector $\boldsymbol{v}$ is along the x-axis, so that $xv = \boldsymbol{r}\,.\,\boldsymbol{v}$, where $\boldsymbol{r} = (x, y, z)$. We use the expression (where $\gamma = \gamma(v)$)

$$\gamma x = (\gamma-1)x+x$$

to rewrite the space component of the right-hand side of (4.20) as

$$(\gamma(x-vt), y, z) = ((\gamma-1)x, 0, 0)+(x, y, z)-(\gamma vt, 0, 0) = (\gamma-1)\frac{\boldsymbol{v}\,.\,(\boldsymbol{v}\,.\,\boldsymbol{r})}{v^2}+\boldsymbol{r}-\gamma\boldsymbol{v}t.$$

Similarly, the time component of the right-hand side of (4.20) is $\gamma\{t-(\boldsymbol{r}\,.\,\boldsymbol{v}/c^2)\}$, so that the Lorentz transformation (4.20) may finally be expressed as:

$$(\boldsymbol{r}', t') = \left((\gamma-1)\boldsymbol{v}\frac{(\boldsymbol{r}\,.\,\boldsymbol{v})}{v^2}+\boldsymbol{r}-\gamma\boldsymbol{v}t, \gamma t-\frac{\gamma\boldsymbol{r}\,.\,\boldsymbol{v}}{c^2}\right). \tag{4.21}$$

We now claim that (4.21) will also be valid for any other direction of $\boldsymbol{v}$, and indeed it is a linear transformation of space and time coordinates which preserves the value of $c^2t^2-r^2$ as can be seen by direct computation:

$$\begin{aligned} c^2t'^2-r'^2 &= -\gamma^2v^2t^2-r^2-(\gamma-1)^2\frac{(\boldsymbol{r}\,.\,\boldsymbol{v})^2}{v^2}-2(\gamma-1)\frac{(\boldsymbol{r}\,.\,\boldsymbol{v})^2}{v^2}+2\gamma(\gamma-1)(\boldsymbol{r}\,.\,\boldsymbol{v})t \\ &\quad +2\gamma(\boldsymbol{r}\,.\,\boldsymbol{v})t+c^2\gamma^2t^2+\gamma^2\frac{(\boldsymbol{r}\,.\,\boldsymbol{v})^2}{c^2}-2\gamma^2t(\boldsymbol{r}\,.\,\boldsymbol{v}) \\ &= \gamma^2(c^2-v^2)t^2-r^2 = c^2t^2-r^2. \end{aligned} \tag{4.22}$$

Thus the speed of a light ray is the same as seen in both frames of reference. It is possible to avoid the calculation of (4.22) by appealing directly to the isotropy of space, so that (4.21), being independent of any particular choice of direction of coordinate axes, will be valid for any such choice.

We note that (4.21) is already reasonably complicated. In order to extend it to the case of arbitrary relative orientation of the two frames of reference we must approach the problem by alternative methods. One of the first steps in doing that is to show that the successive application of two Lorentz transformations is itself a Lorentz transformation with velocity given by the law (4.6) of transformation of velocities. This result in any case is of importance in its own right, since it shows that there is consistency between what two relatively moving observers tell a third who is in motion with respect to both of them.

We will only consider the case of motion along the x-axis. Suppose that v_1 is the relative velocity of the frame of reference S_2 with respect to the frame S_1 and v_2 that of the frame S_3 with respect to S_2. Then if coordinates in the three frames have the corresponding subscripts 1, 2, and 3 respectively, the Lorentz transformation between (x_1, t_1), (x_2, t_2), and (x_3, t_3) are (dropping the y- and z-coordinates):

$$x_2 = \gamma_1(x_1 - v_1 t_1), \qquad t_2 = \gamma_1\left(t_1 - \frac{v_1 x_1}{c^2}\right), \tag{4.23}$$

$$x_3 = \gamma_2(x_2 - v_2 t_2), \qquad t_3 = \gamma_2\left(t_2 - \frac{v_2 x_2}{c^2}\right), \tag{4.24}$$

where $\gamma_i = (1 - v_i^2/c^2)^{-\frac{1}{2}}$, $i = 1, 2$. Replacing x_2 and t_2 in (4.24) by their expression in terms of x_1 and t_1 from (4.23) we have

$$x_3 = \gamma_1\gamma_2\left\{x_1 - v_1 t_1 - v_2\left(t_1 - \frac{v_1 x_1}{c^2}\right)\right\} = \gamma_1\gamma_2 \Big/ \left(1 + \frac{v_1 v_2}{c_2}\right)$$

$$\left\{x_1 - t_1(v_1 + v_2) \Big/ \left(1 + \frac{v_1 v_2}{c^2}\right)\right\}, \tag{4.25}$$

$$t_3 = \gamma_1\gamma_2\left\{t_1 - \frac{v_1 x_1}{c^2} - \frac{v_2(x_1 - v_1 t_1)}{c^2}\right\}$$

$$= \gamma_1\gamma_2\left(1 + \frac{v_1 v_2}{c^2}\right)\left\{t_1 - x_1(v_1 + v_2) \Big/ \left(1 + \frac{v_1 v_2}{c^2}\right)\right\}. \tag{4.26}$$

The structure of the right-hand side of (4.25) and (4.26) will be identical to that of (3.17) and (3.18) if the velocity v in those equations is taken to be the relative velocity

$$v = (v_1 + v_2) \Big/ \left(1 + \frac{v_1 v_2}{c_2}\right). \tag{4.27}$$

That is provided we can prove that

$$\gamma = \left(1-\frac{v^2}{c^2}\right)^{-\frac{1}{2}} = \gamma_1\gamma_2\left(1+\frac{v_1 v_2}{c^2}\right), \tag{4.28}$$

where v on the right-hand side of (4.28) is that given by (4.27). That can be shown by the following algebraic steps:

$$\begin{aligned}
\gamma_1\gamma_2\left(1+\frac{v_1v_2}{c^2}\right) &= \left(1+\frac{v_1v_2}{c^2}\right)\left(1-\frac{v_1^2}{c^2}\right)^{-\frac{1}{2}}\left(1-\frac{v_2^2}{c^2}\right)^{-\frac{1}{2}} \\
&= \left(1+\frac{v_1v_2}{c^2}\right)\Big/\left(1+\frac{v_1^2v_2^2}{c^4}+\frac{2v_1v_2}{c^2}-\frac{2v_1v_2}{c^2}-\frac{v_1^2}{c^2}-\frac{v_2^2}{c^2}\right)^{\frac{1}{2}} \\
&= \left(1+\frac{v_1v_2}{c^2}\right)\Big/\left\{\left(1+\frac{v_1v_2}{c^2}\right)^2-\frac{(v_1+v_2)^2}{c^2}\right\}^{\frac{1}{2}} \\
&= \left[1-\left\{\frac{(v_1+v_2)^2}{c^2}\left(1+\frac{v_1v_2}{c^2}\right)^2\right\}\right]^{-\frac{1}{2}} = \left(1-\frac{v^2}{c^2}\right)^{-\frac{1}{2}}.
\end{aligned}$$

This property of consistency between reports from different observers is the one we will use to study the structure of the most general class of Lorentz transformations. That the successive action of two Lorentz transformations is itself a transformation of the same form indicates that the set of all Lorentz transformations form what is called a group, termed the Lorentz group. We will discuss the very important concept of a group and this particular example of it later when we are at a more suitable stage of understanding. To achieve that we must turn to dynamics.

Further reading for Chapter 4

An expanded version of the discussion is contained in Chapter 5 of *Special relativity* by A. P. French, M.I.T. Introductory Physics Series, Nelson, London (1968). See also Chapter 6 of *Principles of relativity physics* by J. L. Anderson, Academic Press, New York (1967).

5. Relativistic dynamics

5.1. Newtonian energy and momentum

NEWTON's laws of motion allow for the construction of three very important quantities which are at the heart of further developments of dynamics. In fact two of these are essential components of the laws themselves, and all three have proved themselves of great value in correlating the motion of bodies at speeds small compared to that of light. We will start by discussing how exactly these play their parts in Newtonian dynamics before we turn to their extension to bodies in motion at high speeds.

The first quantity is that of inertial mass. It is the constant of proportionality m relating force $\boldsymbol{F}$ and acceleration $\ddot{\boldsymbol{r}}$ in Newton's second law of motion,

$$\boldsymbol{F} = m\ddot{\boldsymbol{r}}. \tag{5.1}$$

Given the value of m for a particle then its motion could be determined, in principle, from (5.1) if the force law were known. For then the problem has been reduced to that of solving a second-order differential equation in the variable $\boldsymbol{r}(t)$ as a function of the time t. The value of m is assumed to be independent of the actual motion of the particle provided it preserves its identity. It is thus a permanent label attached to a particle and should have the same value however it is measured.

It is only very infrequently that the second law of motion (5.1) can be solved explicitly for a given force law, especially if the force arises from the mutual interaction of a number of other similar particles through gravity or the Coulomb force of electromagnetism. However, it is always possible to obtain strong conditions on the motion from what are called conservation laws. These relate the values of conserved quantities, defined initially before the particle or particles are affected by the forces, to their values obtained after all interactions have occurred and the particle or particles are once more moving freely.

We appear to be restricted to situations in which at an early enough and a late enough time all the particles of interest are travelling freely, far distant from each other and all other influences. However, the conserved quantities of interest can be defined at all intermediate time as well, so they act as permanent labels even while interaction is occurring.

The two most important labels for us are energy and momentum. They are always conserved, so that their values for a set of particles a long time before the particles scatter from each other is equal to their values long after they have moved apart. Since energy and momentum can be defined at all times, even during collisions, the total energy and momentum of the system must remain equal to their initial and final values. Such constraints allow very interesting general features of the motion to be obtained, without the far more complicated task of calculating it exactly.

The third of the quantities of interest is linear momentum, which is defined as mass × velocity,

$$\boldsymbol{p} = m\dot{\boldsymbol{r}}.$$

Then (5.1) becomes, for the ith particle of a group of such particles,

$$\boldsymbol{F}_i = \dot{\boldsymbol{p}}_i = m_i\ddot{\boldsymbol{r}}_i, \tag{5.2}$$

where m_i is the inertial mass and $\boldsymbol{p}_i$ the momentum of the ith particle at $\boldsymbol{r}_i$ and $\boldsymbol{F}_i$ is the force on it. We can divide the total force $\boldsymbol{F}_i$ on the ith particle into the external force $\boldsymbol{F}_i^{\text{ext}}$ on it due to external fields and the sum of forces $\boldsymbol{F}_{ij}$ due to the jth particle;

$$\boldsymbol{F}_i = \boldsymbol{F}_i^{\text{ext}} + \sum_{j \neq i} \boldsymbol{F}_{ij}. \tag{5.3}$$

If we now sum (5.2) over i and use (5.3) we obtain

$$\frac{\mathrm{d}}{\mathrm{d}t}\sum_i \boldsymbol{p}_i = \sum_i \boldsymbol{F}_i^{\text{ext}} + \sum_{i,j \neq i} \boldsymbol{F}_{ij}. \tag{5.4}$$

Since by Newton's third law

$$\boldsymbol{F}_{ij} + \boldsymbol{F}_{ji} = 0$$

then

$$\dot{\boldsymbol{P}} = \boldsymbol{F}^{\text{ext}}, \tag{5.5}$$

where $\boldsymbol{P} = \sum \boldsymbol{p}_i$ is the total linear momentum and $\boldsymbol{F}^{\text{ext}}$ is the total external force on the system of particles. When there is no net force, the aggregate of particles being isolated, then $\boldsymbol{F}^{\text{ext}} = 0$, so that from (5.5) $\boldsymbol{P}$ is a constant. We thus have the law of conservation of momentum:

$$\boldsymbol{P}_{\text{initial}} = \boldsymbol{P}_{\text{final}}.$$

We can obtain similarly the law of conservation of energy. We consider a single particle moving in a conservative field of force, being one for which the work done on taking a particle round any closed curve is zero. Then there exists a potential function $V(\boldsymbol{r})$ for which

$$\boldsymbol{F} = -\boldsymbol{\nabla} V \tag{5.6}$$

(where $\boldsymbol{\nabla} = \partial/\partial x,\ \partial/\partial y,\ \partial/\partial z$). The work done in moving the particle from

$\boldsymbol{r}_1$ to $\boldsymbol{r}_2$ is then

$$\int_{\boldsymbol{r}_1}^{\boldsymbol{r}_2} \boldsymbol{F} \cdot \mathrm{d}\boldsymbol{r} = -\int_{\boldsymbol{r}_1}^{\boldsymbol{r}_2} \nabla V \cdot \mathrm{d}\boldsymbol{r} = V(\boldsymbol{r}_1) - V(\boldsymbol{r}_2). \tag{5.7}$$

But by (5.1) this is also equal to

$$\int_{\boldsymbol{r}_1}^{\boldsymbol{r}_2} m\ddot{\boldsymbol{r}} \cdot \mathrm{d}\boldsymbol{r} = \int_{\boldsymbol{r}_1}^{\boldsymbol{r}_2} m\ddot{\boldsymbol{r}} \cdot \dot{\boldsymbol{r}}\, \mathrm{d}t = \int_{\boldsymbol{r}_1}^{\boldsymbol{r}_2} \frac{\mathrm{d}}{\mathrm{d}t}(\tfrac{1}{2}m\dot{\boldsymbol{r}}^2)\, \mathrm{d}t$$

$$= \tfrac{1}{2}m\dot{\boldsymbol{r}}_2^2 - \tfrac{1}{2}m\dot{\boldsymbol{r}}_1^2. \tag{5.8}$$

Combining (5.7) and (5.8) we obtain

$$\tfrac{1}{2}m\dot{\boldsymbol{r}}_1^2 + V(\boldsymbol{r}_1) = \tfrac{1}{2}m\dot{\boldsymbol{r}}_2^2 + V(\boldsymbol{r}_2). \tag{5.9}$$

The quantity E is the total energy of the particle, and from (5.9) is a constant of the motion. It is composed of two parts: the potential energy $V(\boldsymbol{r})$ and the kinetic energy $\frac{1}{2}m\dot{\boldsymbol{r}}^2$. As has already been pointed out it is this total quantity, summed over a number of particles if necessary, which along with the conserved linear momentum, plays such a useful role in realistic dynamical problems.

We realized earlier that Newton's laws are not consistent with the constancy of the velocity of light, an experimental fact at the basis of the Lorentz transformation. It might be expected that energy or momentum defined as above might no longer be conserved for particles moving at speeds close to that of light. That is indeed the case; for example, electrons accelerated through different potentials do not have a resulting kinetic energy which increases

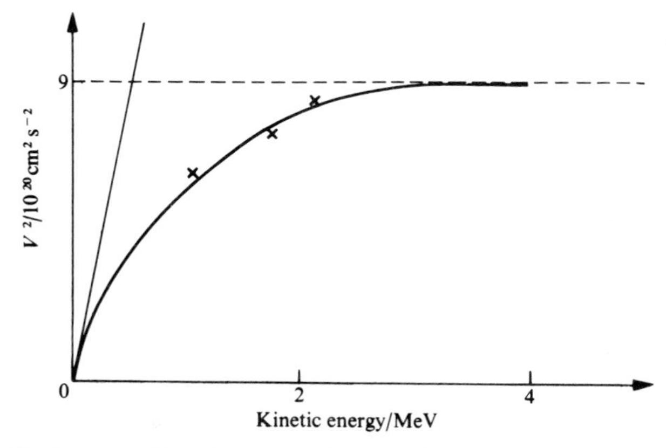

FIG. 16. The increase of kinetic energy, along the abcissa, with velocity for electrons. The experimental points are marked by crosses; the smooth curve is that predicted by (5.13).

quadratically with velocity, as (5.9) indicates it should. This is shown in Fig. 16, where the kinetic energy is found to increase ever higher in spite of the fact that the velocity is apparently limited by the speed of light, as it should be from our earlier discussions.

To understand this situation we have to analyse how momentum and energy are to be defined for particles moving close to that of light in a fashion consistent with the constancy of the speed of light. We will do that by using the rule (4.6) for the transformation of velocities from one frame to another.

5.2. The relativistic case

It is natural to keep as close as we can to the non-relativistic definitions of energy and momentum. We can do this by keeping the momentum $\boldsymbol{p}$ parallel to the velocity $\boldsymbol{v}$ but allowing the mass m to vary as a function of $\boldsymbol{v}$, $m = m(\boldsymbol{v})$. The problem we are faced with is to obtain the explicit dependence of m on $\boldsymbol{v}$. We will do that in a special case and then briefly remark on the extension of the argument to a more general situation.

Consider two equal particles colliding along a straight line. One of the particles is assumed to be at rest in a particular reference frame S, the other to have velocity U before they collide. They then coalesce and the combined object moves with velocity u. The mass of the two initial particles is denoted by $m(U)$ and m_0 and that of the coalesced object by M.

Alternatively, we can consider the collision in a frame in which the two particles collide with equal and opposite speeds, leaving the combined object with mass M_0 at rest. The collisions as seen from the two reference frames is shown in Fig. 17(a) and (b); the frame S′ is evidently moving to the left with

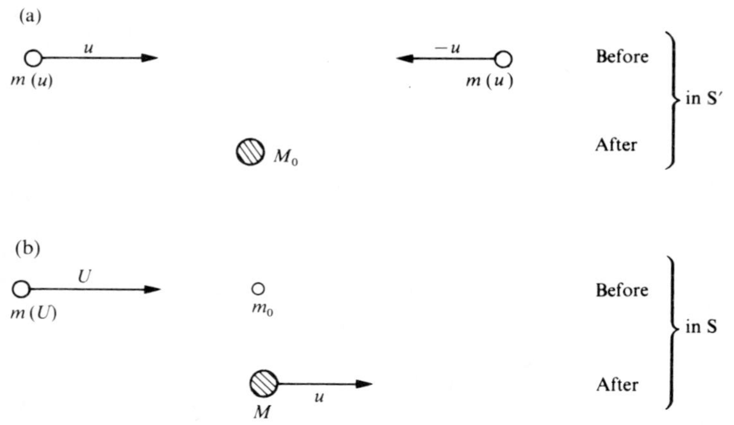

FIG. 17. The collision of two equal particles with resulting coalescence as seen either in (a) their centre-of-mass frame S′ or (b) the rest frame S of one of them.

a velocity u with respect to S, since such a motion applied to S brings the coalesced object to rest.

We assume conservation of both momentum and mass. This latter condition of mass conservation is not one which is immediately obvious, but we will see shortly that it is the extension of energy conservation to fast-moving particles. The initial linear momentum of the two particles, as observed in the frame S, is $m(U)U$ and 0, and finally is Mu; the corresponding masses are $m(U)$, m_0, and M. Thus the two conservation laws are

$$m(U)U = Mu, \tag{5.10}$$

$$m(U)+m_0 = M. \tag{5.11}$$

Dividing each side of (5.10) by the corresponding side of (5.11) to remove M we have

$$m(U)U = u\{m(U)+m_0\}$$

or

$$\frac{m(U)}{m_0} = \frac{u}{(U-u)}. \tag{5.12}$$

We now see that U is the resultant velocity in S′ of the particle moving to the right in the frame S before the collision, this velocity being the result of translating the particle to the right with velocity u in going from the frame S′ to S, the particle already having velocity u to the right in the frame S. In short, U is the resultant of the velocities u and u, so that by the transformation law of velocities

$$U = \frac{2u}{(1+u^2/c^2)}$$

or

$$\frac{2u}{U} = 1+\frac{u^2}{c^2}.$$

By the following steps of algebraic manipulation

$$u^2-\frac{2uc}{U}+c^2 = 0$$

so

$$u = \frac{c^2}{U} \pm \left\{\left(\frac{c^2}{U}\right)^2 - c^2\right\}^{\frac{1}{2}} = \frac{c^2}{U}\left\{1 \pm \left(1-\frac{U^2}{c^2}\right)^{\frac{1}{2}}\right\}$$

$$U-u = \frac{c^2}{U}\left\{\frac{U^2}{c^2}-1+\left(1-\frac{U^2}{c^2}\right)^{\frac{1}{2}}\right\}$$

$$= \frac{c^2}{U}\left(1-\frac{U^2}{c^2}\right)^{\frac{1}{2}}\left\{1-\left(1-\frac{U^2}{c^2}\right)^{\frac{1}{2}}\right\} = \left(1-\frac{U^2}{c^2}\right)^{\frac{1}{2}}u,$$

we obtain that

$$\frac{m(U)}{m_0} = \frac{u}{(U-u)} = \left(1 - \frac{U^2}{c^2}\right)^{-\frac{1}{2}}$$

or

$$m(U) = m_0\left(1 - \frac{U^2}{c^2}\right)^{-\frac{1}{2}}. \tag{5.13}$$

It is this result (5.13) which is the basic one in discussing the dynamics of fast-moving particles. It shows that mass increases with velocity, the behaviour being shown in Fig. 18.

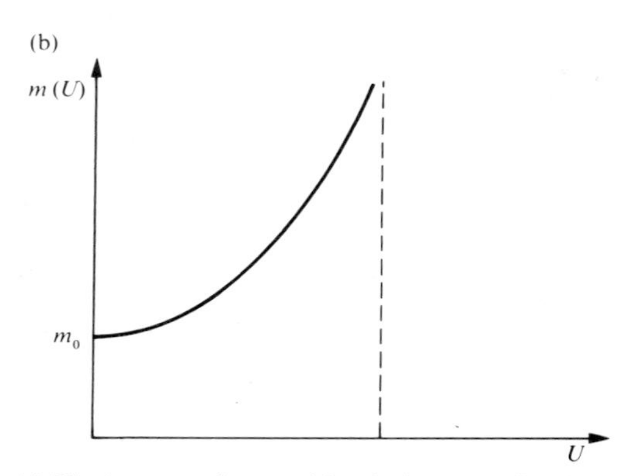

FIG. 18. The increase of mass with velocity as predicted by (5.13).

We have only derived (5.13) by the analysis of a very special dynamical situation, that of the coalescence of two particles. We must be careful to ensure that there are no drastic modifications if other cases are considered, especially if two or more particles are in elastic collisions. This discussion can indeed be given, and it can be shown that (5.13) is indeed the correct definition of the mass of each particle and $Um(U)$ its momentum in the direction of its motion. These are to be used in the equations of momentum and mass conservation for the total dynamical interaction involving the particles.

We will not give all of these further arguments here but will try to sketch out some further grounds for accepting (5.13) as correct. In any case it is not possible to derive (5.13) *a priori*; the only real test is by experiment and has been passed with flying colours.

Let us return, however, to further discussion of (5.13). Since it is usual to think of conservation of energy as well as mass we need to see if energy conservation has still to be considered, beyond the conservation of momentum and mass. Let us evaluate the dependence of $m(U)$ on U for small values of U/c:

$$m(U) \approx m_0 + \frac{\frac{1}{2}m_0 U^2}{c^2} + \cdots. \tag{5.14}$$

We see that if $U \ll c$ then

$$\{m(U) - m_0\}c^2 \approx \tfrac{1}{2}m_0 U^2, \tag{5.15}$$

the right-hand side of (5.15) being the non-relativistic kinetic energy of the particle. Thus we may obtain the energy from $m(U)$ by multiplying by c^2 and substracting the 'rest energy' m_0c^2. Since m_0 is independent of velocity, conservation of the total mass of two colliding particles will lead to the conservation of their total energy; naturally the speeds of all the particles will have to be much less than that of light.

In detail, if the two particles have rest masses m_{10} and m_{20} and initial and final velocities $\boldsymbol{v}_1$, $\boldsymbol{v}'_1$ for the first and $\boldsymbol{v}_2$, $\boldsymbol{v}'_2$ for the second, then the equation of mass conservation is

$$m_{10}\left(1-\frac{v_1^2}{c^2}\right)^{-\frac{1}{2}} + m_{20}\left(1-\frac{v_2^2}{c^2}\right)^{-\frac{1}{2}} = m_{10}\left(1-\frac{v_1'^2}{c^2}\right)^{-\frac{1}{2}} + m_{20}\left(1-\frac{v_2'^2}{c^2}\right)^{-\frac{1}{2}}, \tag{5.16}$$

and when $v_1, v_2, v'_1, v'_2 \ll c$ we have, by expanding each of the terms in (5.16) as in (5.14), multiplying throughout by c^2, and cancelling a common m_{10} and m_{20} on each side,

$$\tfrac{1}{2}m_{10}v_1^2 + \tfrac{1}{2}m_{20}v_2^2 = \tfrac{1}{2}m_{10}v_1'^2 + \tfrac{1}{2}m_{20}v_2'^2, \tag{5.17}$$

which is the usual equation of conservation of energy. In a similar approximation the equation of conservation of momentum becomes

$$m_{10}\boldsymbol{v}_1 + m_{20}\boldsymbol{v}_2 = m_{10}\boldsymbol{v}'_1 + m_{20}\boldsymbol{v}'_2, \tag{5.18}$$

which is the usual one for slow-moving particles.

This fusion of the principles of conservation of mass and energy into the single conservation of the relativistic mass (5.13) is very beautiful. But that is not enough to make formula (5.13) correct. As has already been remarked, it can only be justified by its agreement with experiment. This is especially pressing here since one of the most important properties of $m(U)$, and certainly its most unexpected, is that it increases indefinitely with U as U approaches c, ultimately becoming infinite there. This is very different indeed from the constant mass or the quadratically varying energy of Newtonian physics. Yet it agrees with the experimentally determined variation of energy with speed, as shown in Fig. 16 (p. 47). Although the velocity of the electron is found to be limited by c, its energy can be increased indefinitely. The agreement between the experimental points and the smooth curve following the predictions of (5.13) is evidently very good. This indefinite increase of mass with velocity has been tested even more precisely in the construction of high-energy particle accelerators such as the linear electron accelerator at Stanford, where the energy of the electron is known to increase from its rest value of $\frac{1}{2}$MeV to 20 GeV, an increase by a factor of 4×10^4. A similar increase will occur for the electrons mass. These increases predicted by (5.13) have had to be taken into account very carefully in designing the various machines, since otherwise no particle beam would result; a stringent test indeed!

The result of this discussion is that for a particle moving with velocity $\boldsymbol{v}$ in a given frame of reference S its energy E, momentum $\boldsymbol{p}$, and mass m will be

$$E = mc^2, \qquad \boldsymbol{p} = m\boldsymbol{v}, \qquad m = \gamma m_0, \tag{5.19}$$

where $\gamma = (1 - v^2/c^2)^{-\frac{1}{2}}$. The quantity m_0 is the rest mass of the particle and $m_0 c^2$ is its rest energy.

One of the important consequences of (5.19) is that

$$\left(\frac{E^2}{c^2}\right) - p^2 = m_0^2 c^2. \tag{5.20}$$

In other words the set of four quantities $(E/c, \boldsymbol{p})$ for a particle behave in an identical fashion to the set describing its position at a certain time $(ct, \boldsymbol{r})$. Under Lorentz transformations the lengths $\{(E^2/c^2) - p^2\}$ and $(c^2t^2 - r^2)$ both remain constant whatever the velocity of the particle. This invariance of (5.20) under Lorentz transformations indicates that E/c and $\boldsymbol{p}$ will transform according to the Lorentz transformations (3.17) and (3.18). In a frame S′ moving with velocity $\boldsymbol{u}$ with respect to the frame S, in which the particle has energy E and momentum $\boldsymbol{p}$, the energy E' and momentum $\boldsymbol{p}'$ will be

$$\begin{aligned} \boldsymbol{p}' &= \gamma(u)\{\boldsymbol{p} + (\boldsymbol{u}E|c^2)\}, \\ E' &= \gamma(u)(E + \boldsymbol{p} \cdot \boldsymbol{u}). \end{aligned} \tag{5.21}$$

We have chosen $\boldsymbol{u}$ to be parallel to $\boldsymbol{p}$ for convenience, the more general case being described by (4.21). We have also chosen a positive sign between the first and second term on the right-hand side of (5.21) for the following reason. Let us take S to be the rest frame of the particle, which is that frame in which the particle is at rest, with $\boldsymbol{p} = 0$. Then (5.21) reduces to

$$\boldsymbol{p}' = m(u)\boldsymbol{u}, \qquad E' = m(u)c^2,$$

where

$$m(u) = \gamma(u)m_0, \qquad m_0 = E_0/c^2,$$

and E_0 is the energy of the particle when at rest. But these are the values of the energy, momentum, and mass arrived at in (5.19); a negative sign in the second term on the right-hand side of each of the equations in (5.21) would have led to momentum directed opposite to the particle's velocity.

We conclude that a particle's mass and energy increases indefinitely as its speed approaches that of light. This increase is specified once the rest mass m_0 of a particle is known. This basic parameter is usually given in terms of its energy m_0c^2 in electron volts. The rest masses of the most important elementary particles are shown in Table 4; though they are all extremely small, a gram

TABLE 4

The rest energies of the elementary particles

Particle	Symbol	Rest energy/MeV
Proton	p	938·3
Neutron	n	939·5
Pi-meson	π	140
Mu-meson	μ	106
Electron	e	0·5
Neutrino	ν	0
Photon	γ	0
Graviton	g	0

of hydrogen has a total of 10^{14} J stored in its rest mass. Evidently processes in which rest masses are charged into available energy or vice versa are of great interest. We turn to that now.

5.3. Nuclear binding energies

The most immediate experience of the equivalence between mass and energy expressed by the first of eqns (5.19) is in the nucleus of an atom. Such an object is composed of a certain number of protons and neutrons, particles with about the same mass (see Table 4), the neutron being electrically neutral and the proton having an electrical charge equal in magnitude but opposite in sign to that of the electron. A nucleus will consist of Z protons and N neutrons, where Z is the atomic number, or the correct numerical order in the periodic table of the elements. Since an atom is normally electrically neutral then it will have Z electrons orbiting the nucleus, giving it its chemical properties. The sum $N+Z$ is called the atomic mass number and is usually denoted by A, so $N = A-Z$. If the chemical symbol of the element is X a particular atom of this element can be described by ${}^{A}_{Z}\mathrm{X}$. Thus hydrogen is ${}^{1}_{1}\mathrm{H}$, deuterium is ${}^{2}_{1}\mathrm{H}$, helium ${}^{4}_{2}\mathrm{He}$, and so on.

It is an attractive idea to consider all nuclei as constructed from the fundamental constituents equal to the neutron and proton. Indeed this model is very effective in obtaining many of the properties of nuclei. If we try to calculate the mass of an atom by adding up the masses of its constituents we find there is a small but important discrepancy. The helium atom is composed of two protons, two electrons, and two neutrons, or equivalently of two hydrogen atoms and two neutrons. Thus we might expect the mass M_{He} of the helium atom to be given by

$$M_{\mathrm{He}} = 2(M_{\mathrm{H}}+M_{\mathrm{n}}), \tag{5.22}$$

where M_{H} and M_{n} are the masses of the hydrogen atom and the neutron respectively. A calculation shows (5.22) is not quite correct; if we work in atomic mass units which takes the carbon atom with atomic number 12 to have a weight of 12 mass units (so that 1 mass unit = 931 MeV), we find that

$$M_{\mathrm{H}} = 1{\cdot}007825, \qquad M_{\mathrm{n}} = 1{\cdot}008665,$$

so that

$$M_{\mathrm{He}} = 2\times 2{\cdot}016490 = 4{\cdot}032980.$$

The measured mass of $^{4}_{2}\mathrm{He}$ is 4·002603, less than the mass of the electrons, protons, and neutrons which constitute it by about 0·03 mass units. Similarly, argon 40 ($^{40}_{18}\mathrm{Ar}$) has 18 protons and electrons and 22 neutrons, whose total mass is equal to

$$18\times 1{\cdot}007825+22\times 1{\cdot}008665 = 40{\cdot}331480,$$

whereas the experimental value for the mass of $^{40}_{18}\mathrm{Ar}$ is 39·962384, about 0·37 mass units less than the expected value.

It is found in general that there is always a mass defect $\Delta M_{Z,A}$ for the difference

$$\Delta M_{Z,A} = \{ZM_{\mathrm{H}}+(A-Z)M_{\mathrm{n}}\}-M_{Z,A} \tag{5.23}$$

between the mass of the constituent neutrons and hydrogen atoms and the actual mass $M_{Z,A}$ of an atom. The question is where this mass defect goes to on the formation of an atom from its constituents. The answer is that it is emitted as energy, the amount of energy being called the binding energy of the nucleus (that of the electrons being negligible in comparison). The variation of the binding energy per nucleon is shown in Fig. 19; its value is roughly constant (except for very light nuclei) at about 8·4 MeV. The nuclei at the position of the maximum of this curve are most stable, since more energy is needed to disrupt them into their constituents than for nuclei at either end of the curve.

The validity of this interpretation of the mass defect as binding energy is shown very clearly by radioactive decays. Nuclei can sometimes exist for periods of time but are ultimately unstable to partial break-up. A particularly

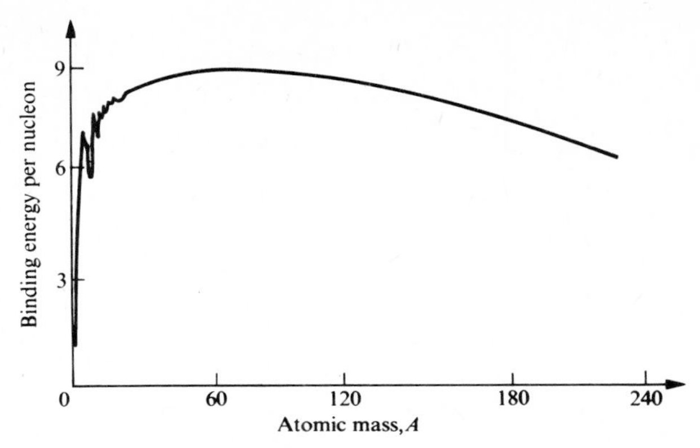

FIG. 19. The binding energy per nucleon for various atoms.

simple case is that of the free neutron which decays in about 12 minutes into a proton, an electron, and a neutrino with no mass or charge. From Table 4 (p. 53) the difference in rest energy between a neutron and a hydrogen atom is 0·783 MeV, so that the decay

$$n \rightarrow p + e^{-} + \nu \tag{5.24}$$

can occur. It is possible to measure the maximum energy the electron can carry, and it is found to agree exactly with its expected value. Similarly many other tests of the relation have been performed in radioactive decays, with no known violation.

We remark finally in this context that the amount of energy available in matter is very large. Thus if $\frac{1}{100}$ gm of matter (the size of a pinhead) were completely annihilated the total energy released would be, in the SI system of units,

$$E = 0{\cdot}01 \times 10^{-3} \times (3 \times 10^{8})^{2} = 9 \times 10^{11}\ \mathrm{J}$$

(where $c = 3 \times 10^{8}\ \mathrm{m\ s^{-1}}$). This amount of energy could place into orbit around the sun a satellite of several metric tonnes weight.

5.4. Particle creation

If particles are accelerated fast enough it is found that they can create new particles by collision. Thus a fast-moving proton striking a stationary one can create a positively charged pi-meson (π^{+}) of rest energy 140 MeV,

$$p + p \rightarrow p + n + \pi^{+}. \tag{5.25}$$

It is clear that reaction (5.25) cannot occur at very low energies. We would expect that since the proton and neutron have about the same mass the incoming proton would need at least 140 MeV extra energy to produce a pi-meson. In actuality this is considerably too small, as we shall see.

We can consider the process in the centre-of-mass frame of the colliding protons, as shown in Fig. 20(a), in which each proton approaches the other with equal but opposite velocity. The total momenta of the two protons is thus initially zero, so that their centre of mass is stationary, hence the epithet 'centre-of-mass frame'.

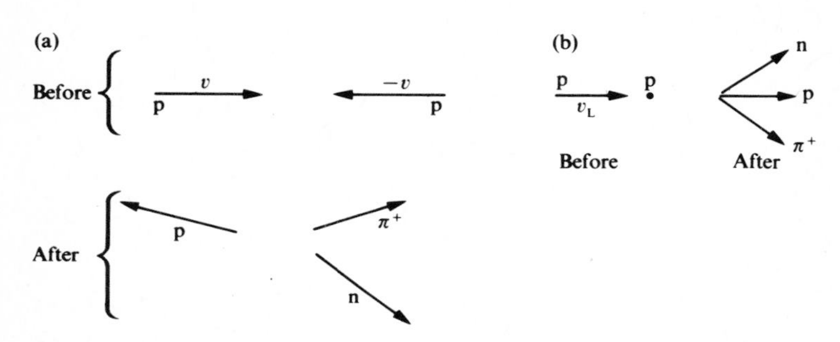

FIG. 20. The creation of a pi-meson in proton–proton collisions as seen (a) in the centre-of-mass frame of the incoming particles, (b) in the laboratory frame of one of the protons.

If the velocities of the protons are $\pm v$ then their total energy is $2m_N\gamma(v)c^2$, where m_Nc^2 is the proton rest energy. Let us neglect the difference between the proton rest mass and neutron rest mass. If the final proton, neutron, and pi-meson can only just be created and so are at rest, their total energy will be $(2m_N+m_\pi)c^2$, where $m_\pi c^2$ is the pi-meson rest energy. The equation of energy conservation is thus

$$2m_N\gamma(v)c^2 = (2m_N+m_\pi)c^2.$$

We obtain

$$\gamma(v) = 1+(m_\pi/2m_N). \tag{5.26}$$

If we use the approximate value $m_\pi/m_N = \frac{1}{7}$ we may solve (5.26) for v/c, to obtain

$$v/c \sim 0{\cdot}37.$$

We may also consider the reaction (5.25) in the laboratory frame, shown in Fig. 20(b), in which the laboratory proton is at rest and the incoming proton has velocity v_L. The laboratory frame is obtained from the centre-of-mass frame by moving the latter to the right with velocity v. The velocity v_L is thus the transform by an additional velocity v of the velocity v of the rightward-moving particle in the centre-of-mass frame,

$$v_{\mathrm{L}} = \frac{2v}{(1+v^2/c^2)} \sim 0{\cdot}65c. \tag{5.27}$$

Then $\gamma(v_{\mathrm{L}}) \sim 1{\cdot}31$, so that the kinetic energy of the bombarding proton in the laboratory frame is

$$\{\gamma(v_{\mathrm{L}})-1\}m_{\mathrm{N}}c^2 \sim 0{\cdot}31m_{\mathrm{N}}c^2 \sim 290 \text{ MeV}.$$

In other words, the additional kinetic energy, measured in the laboratory frame, required to produce a pi-meson is a little more than twice the rest energy of the pi-meson, certainly more than naïvely expected. This result has been fully borne out by experiment.

A similar situation is involved in the production of the anti-proton $\bar{\mathrm{p}}$, a particle identical to the proton except for its opposite charge. It can be produced by the reaction

$$\mathrm{p}+\mathrm{p} \rightarrow \mathrm{p}+\mathrm{p}+\mathrm{p}+\bar{\mathrm{p}}. \tag{5.28}$$

In the centre of mass of the incoming protons, moving with velocity v, the initial energy is $2m_{\mathrm{N}}\gamma(v)c^2$, and the final energy, assuming the three protons and the single anti-proton are only just created at rest, is $4m_{\mathrm{N}}c^2$. Energy conservation thus requires

$$2m_{\mathrm{N}}\gamma(v)c^2 = 4m_{\mathrm{N}}c^2$$

or

$$\frac{v}{c} = \frac{\sqrt{3}}{2}.$$

Then in the laboratory frame, in which one initial proton is at rest and the other is moving with velocity v_{L}, we have as before

$$v_{\mathrm{L}} = \frac{2v}{(1+v^2/c^2)} = \frac{4\sqrt{3}c}{7}, \qquad \gamma_{\mathrm{L}} \sim 7.$$

The kinetic energy needed to create the proton–anti-proton pair is thus $(\gamma_{\mathrm{L}}-1)m_{\mathrm{N}}c^2 \sim 5{\cdot}6$ GeV, about 3 times the rest energy of the pair. One of the reasons the Bevatron accelerator was built to give protons an energy of 6 GeV was to create the anti-proton in this manner; it was produced in this way in 1965.

We can consider also the creation of an electron–positron pair from the photon γ, the particle of light. Thus, in the reaction

$$\gamma \rightarrow \mathrm{e}^{+}+\mathrm{e}^{-}, \tag{5.29}$$

if we are in the centre of mass of the electron and positron so they have equal and opposite momentum, in order for them to be produced at rest the photon energy must be $E = 2m_e c^2 \sim 1$ MeV. In other words, 1 MeV photons will just be able to produce electron–positron pairs. Again this is fully borne out by experiment. We remark finally that for the process (5.29) to proceed it is necessary that material be present to allow momentum conservation to be satisfied. For from (5.20) when $m_0 = 0$, as it is for a photon, then $\mathrm{p} = E/c$. The final electron-positron pair has total momentum zero, so that the non-zero incoming photon momentum E/c must be 'handed on', as it can be if, for example, a nucleus is involved in conjunction with the reaction (5.29).

Further reading for Chapter 5

Nuclear binding energies are discussed in more detail by T. A. Littlefield and N. Thorley in Chapter 14 of *Atomic and nuclear physics*, Van Nostrand, New York (1968).

For particle creation and annihilation see H. H. Heckman and P. W. Starring, *Nuclear physics and the fundamental particles*, Holt, Rinehart, and Winston, New York (1967), together with the more complete discussion of the high-energy physics aspects in R. Hagedorn's *Relativistic kinematics*, Benjamin, New York (1963).

6. Formal relativistic dynamics

6.1. Introduction

WE have already used a considerable amount of space discussing how Lorentz transformations alter the coordinate labels of events. We did that in detail, however, only for the special case of transformations parallel to the x-axis. In order to develop a framework large enough to construct a theory of dynamics consistent with the constancy of the velocity of light we must now extend our discussion of Lorentz transformations to the most general case.

A Lorentz transformation along the x-axis preserves the quantity

$$s_{12}^2 = c^2(t_1-t_2)^2-(\boldsymbol{r}_1-\boldsymbol{r}_2)^2 \tag{6.1}$$

determined from two events $(\boldsymbol{r}, t_1)$, $(\boldsymbol{r}_2, t_2)$. The x-axis is not singled out in (6.1). If we are considering a Lorentz transformation along some other direction than the x-axis we may rotate the three (x, y, z)-axes so that the new x-axis is now moving along the direction of the Lorentz transformation. Since the quantity $(\boldsymbol{r}_1-\boldsymbol{r}_2)^2$ will be unaltered by a rotation of the axes in space then s_{12}^2 of (6.1) should remain unchanged under the Lorentz transformation. In other words, for any two events the quantity s_{12}^2 is the same if measured in a given reference frame S or in any other frame S′ moving at constant speed with respect to S.

We may reverse this property to define a Lorentz transformation of the coordinate labels of an event E; it is any linear transformation of these labels $(\boldsymbol{r}, t)$ which preserves the quantity s_{12}^2 for any pair of events $(\boldsymbol{r}_1, t_1)$ and $(\boldsymbol{r}_2, t_2)$. Any such transformation preserves rectilinear motion by its linearity and, further, respects the constant velocity of light. This is so since a spherical wave of light expanding from any point in the frame S, with event labels $(\boldsymbol{r}_0, t_0)$, is transformed into a similar sphere of light, expanding with the same velocity, as observed in the frame S′ with event labels $(\boldsymbol{r}', t')$ which are the Lorentz transforms of the labels $(\boldsymbol{r}, t)$ of events in S.

A Lorentz transformation is thus a linear transformation

$$(\boldsymbol{r}, t) \to (\boldsymbol{r}', t') \tag{6.2}$$

which is such that for any pair of labels $(\boldsymbol{r}_1, t_1)$, $(\boldsymbol{r}_2, t_2)$ and their transformed values $(\boldsymbol{r}'_1, t'_1)$, $(\boldsymbol{r}'_2, t')$ we have

$$c^2(t_1-t_2)^2-(\boldsymbol{r}_1-\boldsymbol{r}_2)^2 = c^2(t'_1-t'_2)^2-(\boldsymbol{r}'_1-\boldsymbol{r}'_2)^2. \tag{6.3}$$

To develop the conditions (6.2), (6.3) we will now introduce a four-vector notation. This avoids having to treat space and time coordinates separately, giving very compact and elegant expressions for what would otherwise be prohibitively lengthy equations.

The set of four quantities $(ct, \boldsymbol{r})$ will be denoted by the four-vector $\boldsymbol{x}$ with components $(x^0, x^1, x^2, x^3) = (ct, x, y, z)$. A general component of $\boldsymbol{x}$ will be denoted by x^μ, where μ can take any of the values 0, 1, 2, 3. Then a general linear transformation on x^μ can be written

$$x^\mu \rightarrow x'^\mu = \sum_{\nu=0}^{3} \Lambda^\mu_\nu x^\nu, \tag{6.4}$$

where the set of 16 quantities $\Lambda^\mu_\nu(\mu, \nu = 0, 1, 2, 3)$ for a matrix which we will denote by Λ. Then (6.4) can be expressed in matrix notation as $\boldsymbol{x}' = \Lambda\boldsymbol{x}$ (matrix techniques are described briefly in the Appendix). It is useful also to drop the summation symbol $\sum_{\nu=0}^{3}$ in (6.4) and use instead the summation convention. This is that all twice-repeated indices in any equation are to be summed over. Thus the right-hand side of (6.4) is written more simply as $\Lambda^\mu_\nu x^\nu$ using this convention.

To obtain the conditions on Λ in order that (6.3) be true we denote the difference four-vector $(c(t_1 - t_2), (\boldsymbol{r}_1 - \boldsymbol{r}_2))$ by $\boldsymbol{y}$. Then (6.3) is the same as

$$y_0^2 - \boldsymbol{y}^2 = y_0'^2 - \boldsymbol{y}'^2. \tag{6.5}$$

Let us introduce the matrix g with elements $g_{00} = 1$, $g_{ii} = -1$, and $g_{\mu\nu} = 0$ otherwise, so

$$g = \begin{bmatrix} 1 & 0 & 0 & 0 \\ 0 & -1 & 0 & 0 \\ 0 & 0 & -1 & 0 \\ 0 & 0 & 0 & -1 \end{bmatrix}.$$

Then

$$y_0^2 - \boldsymbol{y}^2 = y^\mu g_{\mu\nu} y^\nu, \tag{6.6}$$

where the summation convention occurs twice in (6.6), once for the index μ and once for ν. We may rewrite (6.5), using (6.6), as

$$y^\mu g_{\mu\nu} y^\nu = y'^\mu g_{\mu\nu} y'^\nu$$

and replacing $\boldsymbol{y}'$ by $\boldsymbol{y}$ by means of (6.4),

$$y^\mu g_{\mu\nu} y^\nu = \Lambda^\mu_a y^a g_{\mu\nu} \Lambda^\nu_b y^b. \tag{6.7}$$

By interchanging the indices μ and a, ν and b in (6.7), since these are to be summed over so that their names are unimportant, we obtain

$$y^\mu (g_{\mu\nu} - \Lambda^a_\mu g_{ab} \Lambda^b_\nu) y^\nu = 0. \tag{6.8}$$

Since (6.8) must be true for any $\boldsymbol{y}$ it is not difficult to convince oneself that (6.8) can only be satisfied in this way provided each of the quantities inside the bracket is zero,

$$g_{\mu\nu} = \Lambda^a_\mu g_{ab} \Lambda^b_\nu \tag{6.9}$$

for $\mu, \nu = 0, 1, 2, 3$. The condition (6.9) may be written more compactly in matrix notation,

$$g = \Lambda^{\mathrm{T}} g \Lambda, \tag{6.10}$$

where the transpose matrix Λ^{T} is defined by $\Lambda^{\mathrm{T}a}_b = \Lambda^b_a$. Any matrix Λ satisfying (6.10) thus defines a Lorentz transformation by (6.4).

The simplest Lorentz transformation is that with which we started, along the x-axis. It is given, by (3.17) and (3.18), to be

$$x' = \gamma\left(x - \frac{vt}{c}\right), \qquad ct' = \gamma\left(ct - \frac{vx}{c}\right).$$

The matrix Λ for this is then

$$\Lambda = \begin{bmatrix} \gamma & 0 & 0 & -\gamma v/c \\ 0 & 1 & 0 & 0 \\ 0 & 0 & 1 & 0 \\ -\gamma v/c & 0 & 0 & \gamma \end{bmatrix}. \tag{6.11}$$

It is straightforward to show that Λ given by (6.11) satisfies (6.10), since by direct calculation

$$\Lambda^{\mathrm{T}} g \Lambda = \begin{bmatrix} \gamma & 0 & 0 & -\gamma v/c \\ 0 & 1 & 0 & 0 \\ 0 & 0 & 1 & 0 \\ -\gamma v/c & 0 & 0 & \gamma \end{bmatrix} \begin{bmatrix} 1 & 0 & 0 & 0 \\ 0 & -1 & 0 & 0 \\ 0 & 0 & -1 & 0 \\ 0 & 0 & 0 & -1 \end{bmatrix} \begin{bmatrix} \gamma & 0 & 0 & -\gamma v/c \\ 0 & 1 & 0 & 0 \\ 0 & 0 & 1 & 0 \\ -\gamma v/c & 0 & 0 & \gamma \end{bmatrix}$$

$$= g.$$

6.2. Why special relativity?

We have occasionally used the phrase 'special relativity', but up to now in this book have only given a very brief explanation of what the phrase means.

We are now in a position to explain the phrase in a fashion which allows us to implement it; in this section we will attempt to do that.

It is the principle of relativity which is at the basis of special relativity. This principle states that the world looks essentially the same to each one of a privileged class of observers. More specifically, no one of these observers can be singled out from the others by any observations he makes of the world about him. This means that the laws of dynamics which the various observers use to correlate their experimental results have the same form for all the observers. None of them can then say that his or her understanding of the world is different, in principle, from any of their colleagues. We can then say that their dynamical laws are relativistically covariant.

The principle of relativity has to be made more precise than this to give it any effect. In particular, it is necessary to specify the class of observers to whom it applies. Thus Galilean relativity considers the set of observers who are in uniform relative motion with respect to each other and each of whom is an inertial observer, so that the law of inertia—freely moving particles travel along straight paths—is valid for them. We saw earlier that Newton's laws of motion preserve their form under Galilean transformations. Thus Newtonian mechanics satisfies the principle of Galilean relativity for the class of inertial observers.

The problem we are now faced with is to construct a mechanics for matter which is consistent with the experimental fact of the constancy of light. We realized earlier that Galilean relativity is inconsistent with this constant velocity. Indeed, this latter can only be preserved if Galilean transformations are replaced by Lorentz transformations. Thus it would seem more hopeful to consider the class of observers whose coordinate labels are related by Lorentz transformations. Thus we will search for a set of form-covarient dynamical laws to be used by observers related by Lorentz transformations. In other words, the form of the dynamical law for a given observer must be preserved when it is translated to another observer's frame of reference by a suitable Lorentz transformation.

The law of inertia is extremely well borne out by experiment, both on slowly moving particles of macroscopic size and on elementary particles moving at close to the speed of light. Since the Lorentz transformation is linear on the coordinate labels it does not modify rectilinear motion. Thus we may restrict our discussion to the class of inertial frames of reference. This is a very important simplification, since otherwise we expect dynamical laws to contain extra terms due to acceleration against the inertial frame of the fixed stars.

The principle of special relativity is the requirement of relativity restricted to the class of inertial observers whose coordinate labels are related by Lorentz transformations. It is this principle which guided Einstein in 1905 to set up a dynamics of mass points and of electrodynamics. The extension of this principle to all of the forces of nature, in particular that of gravity, was achieved later, though it has not yet been obtained satisfactorily.

There are two basic difficulties in developing the theory of special relativity. First of all, it is clearly necessary to alter Newton's laws since they only satisfy Galilean relativity and not special relativity. Secondly, the forces met with in dynamics are usually transmitted instantaneously. Simultaneity cannot be defined in a Lorentz covariant fashion—two events may only be simultaneous in one special frame of reference and cannot be in all others. Thus the forces of nature have to be modified so as their effect is propagated with a finite speed. We will turn to discuss these two difficulties in the next two sections.

6.3. Dynamics of mass points

In the last chapter we considered various collision processes between particles. We are able to give a Lorentz covariant discussion there since the forces between the colliding particles only acted instantaneously at the point of collision in space; they did not have to be propagated at infinite speed from one point of space to another. When we turn to discuss how Newton's laws may be modified for a single particle we have to take more serious thought.

The first feature we must modify in Newtonian mechanics is that of mass. We obtained the mass–velocity relation in § 5.2,

$$m(\boldsymbol{v}) = \gamma(\boldsymbol{v})m_0.$$

The momentum is $\boldsymbol{v}m(\boldsymbol{v})$, so that a first guess to the modification of Newton's second law of motion is that the force $\boldsymbol{F}$ on the particle when in motion with velocity $\boldsymbol{v}$ is given by

$$\boldsymbol{F} = \frac{\mathrm{d}}{\mathrm{d}t}\{m(\boldsymbol{v})\boldsymbol{v}\}. \tag{6.12}$$

The right-hand side of (6.12) is

$$m(\boldsymbol{v})\dot{\boldsymbol{v}} + \frac{\boldsymbol{v}m_0(\boldsymbol{v}\,.\,\dot{\boldsymbol{v}})}{c^2(1-v^2/c^2)^{\frac{3}{2}}}$$

which is a complicated expression which is only parallel to the acceleration if $\boldsymbol{v}$ and $\dot{\boldsymbol{v}}$ are orthogonal or parallel. In the former case

$$\boldsymbol{F} = m(\boldsymbol{v})\dot{\boldsymbol{v}}, \tag{6.13}$$

and in the latter

$$\boldsymbol{F} = \frac{m(\boldsymbol{v})\dot{\boldsymbol{v}}}{(1-v^2/c^2)}. \tag{6.14}$$

It is natural to call the mass in (6.13) the transverse mass and that in (6.14) the longitudinal mass.

The force $\boldsymbol{F}$ defined by (6.12) has no simple transformation property under a Lorentz transformation. It can certainly be used as the definition of force, in spite of this complexity, but it is more convenient to aim for a prescription of force which transforms simply under a Lorentz transformation. Since it contains at least three components, the smallest object defined in a Lorentz covarient fashion is a four-vector (F^0, F', F^2, F^3) which transforms like the four-vector x^μ as specified by (6.4) under a Lorentz transformation.

In order to introduce such a four-vector force we must start by defining a velocity four-vector extending the Newtonian velocity $\dot{\boldsymbol{r}}$. This can be done by means of the proper time which is obtained from the invariant of (6.1). For small differences of space and time labels $\mathrm{d}\boldsymbol{r}$ and $\mathrm{d}t$ between two nearby events we define (following (6.1)) their invariant separation $\mathrm{d}s$ by

$$c^2(\mathrm{d}s)^2 = c^2(\mathrm{d}t)^2 - (\mathrm{d}\boldsymbol{r})^2. \tag{6.15}$$

Then the quantity $\mathrm{d}x^\mu/\mathrm{d}s$ is a four-vector transforming under a Lorentz transformation exactly as x^μ does in (6.4). It is this four-vector quantity which replaces $\dot{\boldsymbol{r}}$ in the relativistic mechanics of point particles. We may relate these two quantities by using $\mathrm{d}s = \mathrm{d}t(1-v^2/c^2)^{\frac{1}{2}}$, so that for $i = 1, 2, 3$

$$\frac{\mathrm{d}x^i}{\mathrm{d}s} = \frac{\mathrm{d}x^i}{\mathrm{d}t}\frac{\mathrm{d}t}{\mathrm{d}s} = \gamma(\boldsymbol{v})\dot{x}^i. \tag{6.16}$$

If the particle is moving slowly that $\gamma(\boldsymbol{v}) \approx 1$, so $\mathrm{d}x^i/\mathrm{d}s = \dot{x}^i$, the Newtonian velocity. We note that $\mathrm{d}t/\mathrm{d}s = \gamma(\boldsymbol{v})$, so that if m_0 is a constant,

$$m_0\frac{\mathrm{d}x^i}{\mathrm{d}s} = \gamma(\boldsymbol{v})m_0\dot{x}^i = p^i,$$

$$m_0\frac{\mathrm{d}t}{\mathrm{d}s} = \gamma(\boldsymbol{v})m_0 = m(\boldsymbol{v}) = E/c^2.$$

Thus the four-vector

$$m_0\frac{\mathrm{d}x^\mu}{\mathrm{d}s} = \left(m_0\frac{\mathrm{d}x^i}{\mathrm{d}s}, m_0 c\frac{\mathrm{d}t}{\mathrm{d}s}\right) = \left(p^i, \frac{E}{c}\right). \tag{6.17}$$

We may now define the four-vector force F_μ on a point particle of energy–momentum four-vector $p^\mu = (p^i, E/c)$ to be

$$F^\mu = \frac{\mathrm{d}p^\mu}{\mathrm{d}s}. \tag{6.18}$$

The space components of (6.18) are

$$F^i = \frac{\mathrm{d}p^i}{\mathrm{d}s} = \dot{p}^i\gamma(\boldsymbol{v}) \approx \dot{p}^i, \tag{6.19}$$

where the last step of (6.19) is true for slowly moving particles for which $\gamma(\boldsymbol{v}) \approx 1$. Thus the force law (6.18) is a special relativistically covariant extension of Newton's second law of motion.

We note that we have a further equation in (6.18) beyond the three of Newton's second law. This equation is

$$F^0 = \frac{1}{c}\frac{\mathrm{d}E}{\mathrm{d}s} = \gamma(\boldsymbol{v})\frac{\dot{E}}{c}. \tag{6.20}$$

For slowly moving particles the right-hand side of (6.20) is proportional to the time rate of change of energy, so that F^0 on the left-hand side denotes the power being supplied to the particle by the forces on it.

We have thus resolved the first difficulty of relativistic mechanics, that of setting up covariant equations of motion. The only problem now is to obtain a covariant force F^μ by other means, so that eqn (6.18) has dynamical content, and is not merely a definition of F^μ. We turn to that now for charged particles.

6.4. Electrodynamics

Both the section and the next chapter are somewhat more sophisticated than those up to now, especially at a mathematical level. Those who wish to do so can continue directly to the last chapter without loss of basic understanding. However both items are of utmost importance to those who wish to study modern theoretical physics further.

The force holding together the atoms or molecules of solids, and even the electrons to nuclei, is that of electricity and magnetism. This force is correctly described by Maxwell's equations, one of the most important set of equations in physics. These equations are especially relevant since they also describe the propagation of light, upon which special relativity is based.

We will start with a discussion of Maxwell's equations which determine the electric field $\boldsymbol{E}$ and the magnetic field $\boldsymbol{H}$ generated by a distribution of electric charge of density ρ and current density $\boldsymbol{j}$. Maxwell's equations are (neglecting polarization effects in the material through which the electric and magnetic fields are moving)

$$\begin{aligned} \nabla \cdot \boldsymbol{E} &= \frac{1}{\varepsilon_0}\rho, & \nabla \cdot \boldsymbol{H} &= 0, \\ \nabla \times \boldsymbol{E} + \mu_0 \dot{\boldsymbol{H}} &= 0, & \nabla \times H - \varepsilon_0 \dot{E} &= j \end{aligned} \tag{6.21}$$

where S.I. units are used with ε_0 and μ_0 being the dielectric constant and magnetic permeability of free space. The first two equations state that ρ is the source density for $\boldsymbol{E}$ and that there is no source of $\boldsymbol{H}$ (assuming there are no magnetic poles). The remaining equations are Faraday's law of induction and the generalization of Ampere's law.

It is to be hoped that eqns (6.21) are already Lorentz covariant, since the propagation of light in free space with a constant velocity c can be predicted from them. This can be seen by taking the curl of either the third or fourth equations (with $\boldsymbol{j} = 0$), and using the other equation and the first two equations (with $\rho = 0$),

$$\nabla \times (\nabla \times \boldsymbol{E}) = \nabla(\nabla \cdot \boldsymbol{E}) - \nabla^2 \boldsymbol{E} = -\nabla^2 \boldsymbol{E} = -\mu_0 \nabla \times \dot{\boldsymbol{H}} = -\mu_0 \varepsilon_0 \ddot{\boldsymbol{E}}$$

or

$$\frac{1}{c^2}\ddot{\boldsymbol{E}} - \nabla^2 \boldsymbol{E} = 0, \qquad (6.22)$$

where $c = (\mu_0 \varepsilon_0)^{-\frac{1}{2}} = 3 \times 10^{10}$ cm sec^{-1}.

The solutions of (6.22) are generally of form $\boldsymbol{E} \times \boldsymbol{E}_0 f(\boldsymbol{r} \cdot \boldsymbol{n} - ct)$, where $\boldsymbol{E}_0$ and $\boldsymbol{n}$ are constant vectors and f is any twice-differentiable function, as can be proved by direct substitution. Such a solution represents a wave travelling in the direction of $\boldsymbol{n}$ with velocity $\boldsymbol{c}$, so the prediction is correct; the velocity $(\mu_0 \varepsilon_0)^{-\frac{1}{2}}$ is in close numerical agreement with that of light discussed earlier.

To show the Lorentz covariance of Maxwell's equations (6.21) we have to express them in a covariant form. We can do that if we collect $\boldsymbol{E}$ and $\boldsymbol{H}$ together; there are then six quantities in all. They cannot be the components of a four-vector but we can go one step higher to a product of two four-vectors. If the vectors were $\boldsymbol{x}$ and $\boldsymbol{y}$ we could form the two-index quantity

$$T^{\mu\nu} = x^\mu y^\nu \qquad (\mu, \nu = 0, 1, 2, 3). \qquad (6.23)$$

Under a simultaneous Lorentz transformation on $\boldsymbol{x}$ and $\boldsymbol{y}$ the sixteen quantities $T^{\mu\nu}$ transform as

$$T^{\mu\nu} \to T^{\mu\nu\prime} = \Lambda^\mu_a \Lambda^\nu_b T^{ab}. \qquad (6.24)$$

We note here that we are using the summation convention introduced in section (6.1); thus (6.24) written out in full would read

$$T^{\mu\nu} \to T^{\mu\nu\prime} = \Lambda^\mu_0 \Lambda^\nu_0 T^{00} + \Lambda^\mu_1 \Lambda^\nu_0 T^{10} + \Lambda^\mu_1 \Lambda^\nu_1 T^{11} + \cdots.$$

We can generalize from the case of (6.23) to a general two-index quantity $T^{\mu\nu}$, with sixteen independent components, and define its Lorentz transformation by (6.24). Such a quantity $T^{\mu\nu}$ is called a tensor.

We might hope that we can express $\boldsymbol{E}$ and $\boldsymbol{H}$ as the components of a suitable tensor which we will denote by $F^{\mu\nu}$. It is necessary to remove the further ten of the sixteen components of $F^{\mu\nu}$, since we only require six quantities in all. This can be done easily by requiring $F^{\mu\nu}$ to be anti-symmetric in its indices,

$$F^{\mu\nu} = -F^{\nu\mu}. \qquad (6.25)$$

Then there will only be the six independent quantities F^{01}, F^{02}, F^{03}, F^{12}, F^{23}, F^{31}, all of the other components of $F^{\mu\nu}$ either being expressible in terms of them, as in the case of F^{10}, F^{20}, F^{30}, F^{21}, F^{32}, F^{13}, or zero, as in the case of F^{00}, F^{11}, F^{22}, F^{33}. We make the identification

$$E_1 = (\varepsilon_0 c)^{-1}F^{01}, \qquad E_2 = (\varepsilon_0 c)^{-1}F^{02}, \qquad E_3 = (\varepsilon_0 c)^{-1}F^{03}, \qquad H_1 = F^{23},$$
$$H_2 = F^{31}, \qquad H_3 = F^{12}. \tag{6.26}$$

It will turn out that this identification will allow for the most elegant Lorentz-covariant expression for eqns (6.21). Before we can obtain that we have to represent ρ and $\boldsymbol{j}$ Lorentz covariantly. We can evidently combine these into a four-vector j^μ with $j^0 = c\rho$, and this we will do; the four vector character of j^μ follows from the fact that j^μ is ρ times the velocity four-vector $\mathrm{d}x^\mu/\mathrm{d}s$ for each charged particle.

To express (6.21) in terms of $F^{\mu\nu}$ and j^μ we evaluate the four-divergence $\sum_{\mu=0}^{3} \partial/\partial x^\mu F^{\mu\nu}$; we will denote $\partial/\partial x^\mu$ by ∂_μ, so that by the summation convention this quantity is $\partial_\mu F^{\mu\nu}$. Its time and x-components, written in full, are

$(\nu = 0)$

$$\sum_{i=1}^{3} \partial_i F^{i0} = -\varepsilon_0 c \boldsymbol{\nabla} \,.\, \boldsymbol{E}, \tag{6.27}$$

$(\nu = 1)$

$$\partial_0 F^{01} + \sum_{i=1}^{3} \partial_i F^{i1} = \varepsilon_0 \dot{E}_1 + \partial_2 F^{21} + \partial_3 F^{31}$$
$$= \varepsilon_0 \dot{E}_1 - \partial_2 H_3 + \partial_3 H_2. \tag{6.28}$$

(plus similar equations for the y- and z-components). By the first equation of (6.21) the right-hand side of (6.27) is $-j^0$, and by the last equation of (6.21) the right-hand side of (6.28) is $-j^1$. Thus (6.27) and (6.28) may be combined to give

$$\partial_\mu F^{\mu\nu} = -j^\nu. \tag{6.29}$$

Evidently the right-hand side of (6.29) transforms under Lorentz transformations as a four vector. We can also prove that the left-hand side does the same. For if $\boldsymbol{x} \to \boldsymbol{x}' = \Lambda \boldsymbol{x}$, then $\partial_\mu \to \partial'_\mu = \partial/\partial x'^\mu = \partial_\nu \,.\, \partial x^\nu/\partial x'^\mu$, and $\boldsymbol{x} = \Lambda^{-1}\boldsymbol{x}'$, so that $\partial x^\nu/\partial x'^\mu = (\Lambda^{-1})^\nu_\mu$. Then with (6.24) we have

$$\partial_\mu F^{\mu\nu} \to \partial'_\mu F'^{\mu\nu} = \partial_a (\Lambda^{-1})^a_\mu \Lambda^\mu_b \Lambda^\nu_c F^{bc} = \partial_a (\Lambda^{-1}\Lambda)^a_b \Lambda^\nu_c F^{bc}$$
$$= \Lambda^\nu_c \partial_a F^{ac},$$

which is just the transformation rule of a four-vector under a Lorentz transformation. So both sides of (6.29) transform identically under a Lorentz transformation, and in a new frame of reference with dashed labels for the variables the equation is identical in form:

$$\partial'_{\mu} F'^{\mu\nu} = -j'^{\mu}. \tag{6.30}$$

We have finally to express the second and third equations of (6.21) in covariant form. This can be done by means of the quantity

$$\partial^{\mu} F^{\nu\mu} + \partial^{\lambda} F^{\mu\nu} + \partial^{\nu} F^{\lambda\mu}, \tag{6.31}$$

where μ, λ, ν are any three of 0, 1, 2, 3. Here we define the process of raising the index μ on ∂_{μ}, by means of the matrix g, as $\partial^{\mu} = g^{\mu\nu}\partial_{\nu}$, where

$$g^{\mu\nu} = (g^{-1})^{\mu\nu} = g_{\mu\nu};$$

thus

$$\partial^1 = -\partial_1, \qquad \partial^2 = -\partial_2, \qquad \partial^3 = -\partial_3, \qquad \partial^0 = \partial_0.$$

We can immediately dismiss the cases in which any pair are equal since the quantity (6.31) is then trivially zero, as, for instance, with $\mu = \nu = 1$, $\lambda = 2$, when it becomes

$$\partial' F^{12} + \partial^2 F^{11} + \partial^1 F^{21} = \partial^1 (F^{12} + F^{21}) = 0,$$

the anti-symmetry expressed by (6.25) being used in the last step. If μ, ν, and λ are all distinct, say $\mu = 1$, $\nu = 2$, $\lambda = 3$, then (6.31) becomes

$$\partial^1 F^{23} + \partial^3 F^{12} + \partial^2 F^{31} = -\partial_1 H_1 - \partial_2 H_2 - \partial_3 H_3 = -\boldsymbol{\nabla} \cdot \boldsymbol{H}, \tag{6.32}$$

while if $\mu = 0$, $\nu = 1$, $\lambda = 2$, say, then (6.31) is

$$\begin{aligned} \partial^0 F^{12} + \partial^2 F^{01} + \partial^1 F^{20} &= \frac{1}{c}\dot{H}_3 - \varepsilon_0 c \partial_2 E_1 + \varepsilon_0 c \partial_1 E_2 \\ &= \left(\frac{1}{c}\dot{\boldsymbol{H}} + \varepsilon_0 c \boldsymbol{\nabla} \times \boldsymbol{E}\right)_3. \end{aligned} \tag{6.33}$$

Since one of λ, μ, ν is 0 or they are some permutation of 1, 2, 3 then (6.32) and (6.33) cover all the independent cases of (6.31). We can re-express the second and third equations of (6.21), using (6.32) and (6.33), as the set of equations

$$\partial^{\mu} F^{\nu\lambda} + \partial^{\lambda} F^{\lambda\nu} + \partial^{\nu} F^{\lambda\mu} = 0. \tag{6.34}$$

We note that the set (6.34) are covariant under Lorentz transformations, the left-hand side being an object transforming under the generalization of the transformation law (6.24) to a three-index tensor $T^{\mu\nu\lambda}$,

$$T^{\mu\nu\lambda} \to T'^{\mu\nu\lambda} = \Lambda^{\mu}_{a}\Lambda^{\nu}_{b}\Lambda^{\lambda}_{c} T^{abc}. \tag{6.35}$$

That (6.35) is correct for the left-hand side of (6.34) follows from the transformation law (6.24) for $F^{\mu\nu}$, $F^{\nu\lambda}$, or $F^{\lambda\mu}$, wherever they appear, and the fact that ∂^μ transforms as a vector,

$$\partial^\mu \to \partial'^\mu = g^{\mu\nu}\partial'\nu = g^{\mu\nu}(\partial x^\lambda/\partial x'^\nu)\partial_\lambda = g^{\mu\nu}(\Lambda^{-1})^\lambda_\nu\partial_\lambda = (\Lambda^{-1})^\lambda_\nu g^{\nu\mu}\partial_\lambda$$

$$= (\Lambda^{-1}g)^{\lambda\mu}\partial_\lambda = (g\Lambda^{\mathrm{T}})^{\lambda\mu}\partial_\lambda = g^{\lambda\nu}\Lambda^\mu_\nu\partial_\lambda = \Lambda^\mu_\nu\partial^\nu. \tag{6.36}$$

We have used in (6.36) that $\Lambda g\Lambda^{\mathrm{T}} = g$, so that $\Lambda^{-1}g = g\Lambda^{\mathrm{T}}$, and that g is symmetric. Evidently (6.36) is the transformation law of a four-vector, which is the reason that we introduced the set of operators ∂^μ with the index raised above. Indeed it is usual to label the set of quantities which transform as x^μ by labels in the same 'upstairs' place that they occur on x^μ. This is similarly true for the higher tensors $T^{\mu\nu}$, $T^{\mu\nu\lambda}$, $T^{\mu\nu\lambda\sigma}$, etc. Quantities which transform in this fashion are termed contravariant; those which transform like ∂_μ,

$$\partial_\mu \to \partial'_\mu = (\Lambda^{-1})^\nu_\mu\partial_\nu$$

are termed covariant and written T_μ, $T_{\mu\nu}$, etc.

Thus we finally have obtained a Lorentz covariant form (6.30), (6.34) for the Maxwell equations. The other expression which needs to be written covariantly is the Lorentz force $\boldsymbol{F} = e(\boldsymbol{E}+1/c\boldsymbol{v}\times\boldsymbol{H})$ on a particle moving with velocity $\boldsymbol{v}$ and charge e in the electric field $\boldsymbol{E}$ and the magnetic field $\boldsymbol{B}$. We can extend this to the case of a continuous distribution of charged matter with charge density ρ and current vector $\boldsymbol{j}$, which will be acted on by a force $\boldsymbol{F}$ per unit volume equal to

$$\boldsymbol{F} = (\rho\boldsymbol{E}+\boldsymbol{j}\times\boldsymbol{B}). \tag{6.37}$$

If we consider the expression $j_\mu F^{\mu\nu}$, where $j_\mu = g_{\mu\nu}j^\nu$, so $j_0 = j^0 = c\rho, j_i = -j^i$, then the first component of $j_\mu F^{\mu\nu}$ is

$$j_\mu F^{\mu 1} = j_0F^{01}+j_kF^{k1} = \varepsilon_0\rho c^2E_1-j_2F^{21}-j_3F^{31}$$

$$= \varepsilon_0c^2\rho E_1+j_2H_3-j_3H_2 = \frac{1}{\mu_0}(\rho\boldsymbol{E}+\boldsymbol{j}\times\boldsymbol{B})_1 = \frac{1}{\mu_0}F^1.$$

Thus the space components of $\mu_0 j_\mu F^{\mu\nu}$ are identical to those of the Lorentz force acting on the charge distribution. There is still the time component of $\mu_0 j_\mu F^{\mu\nu}$, whose value is

$$\mu_0\sum_{k=1}^{3} j_kF^{k0} = -\frac{1}{c}\boldsymbol{j}\,.\,\boldsymbol{E}.$$

But the rate of working per unit volume of the Lorentz force $\boldsymbol{F}$ on the charge distribution is $(1/\rho)\boldsymbol{j}\,.\,\boldsymbol{F}$, the local velocity of the charges being $\boldsymbol{j}/\rho$, and since $\boldsymbol{j}$ is orthogonal to $\boldsymbol{j}\times\boldsymbol{B}$ this, by (6.37), is just equal to $\boldsymbol{j}\,.\,\boldsymbol{E}$. Thus the time component of $\mu_0\boldsymbol{j}_\mu F^{\mu\nu}$ is the rate of work being done by the charge distribution in its motion, divided by c.

The quantity $j_\mu F^{\mu\nu}$ again transforms like a four-vector, using the same argument as for $\partial_\mu F^{\mu\nu}$. Indeed, the contracted tensor $T_\mu^{\mu\nu\lambda\ldots\sigma}$ defined from the tensor $T_\alpha^{\mu\nu\lambda\ldots\sigma}$ by setting μ and α equal and summing over their possible values 0, 1, 2, 3, always transforms like a tensor on the indices $\nu, \lambda, \ldots, \sigma$. This follows from the transformation law for the mixed tensor $T_\mu^{\nu\lambda\ldots\sigma}$,

$$T_\mu^{\nu\lambda\ldots\sigma} \to T_\mu'^{\nu\ldots\sigma} = (\Lambda^{-1})_\mu^a \Lambda_b^\nu \Lambda_c^\lambda \ldots \Lambda_d^\sigma T_a^{bc\ldots d}. \tag{6.38}$$

For then

$$\begin{aligned} T_\mu'^{\mu\lambda\ldots\sigma} &= (\Lambda^{-1})_\mu^a \Lambda_b^\mu \Lambda_c^\lambda \ldots \Lambda_d^\sigma T_a^{bc\ldots d} \\ &= \Lambda_c^\lambda \ldots \Lambda_d^\sigma T_a^{ac\ldots d}, \end{aligned}$$

and there is contravariance in the labels $\lambda \ldots \sigma$. A similar discussion can be given if $T_{\cdots}^{\cdots}$ has several covariant as well as contravariant indices; the contracted tensor $T_{a\lambda\ldots\sigma}^{a\mu\ldots\nu}$ always transforms contravariantly in $\mu \ldots \nu$ and covariantly $\lambda \ldots \sigma$.

We finish this chapter by remarking that the transformation law (6.24) indicates that $\boldsymbol{E}$ and $\boldsymbol{H}$ are mixed up by motion. Suppose that in a particular frame of reference S there is only an electric field $\boldsymbol{E}$ (say that of a point particle). If there is a transformation with velocity $\boldsymbol{v}$ along the x-axis to a frame of reference S′, then the values $\boldsymbol{E}'$ and $\boldsymbol{H}'$ of the electric and magnetic fields in S′ will be, by (6.24),

$$\left.\begin{aligned} E_1' &= F'^{01} = \Lambda_\mu^0 \Lambda_\nu^1 F^{\mu\nu}, \\ E_2' &= F'^{02} = \Lambda_\mu^0 \Lambda_\nu^2 F^{\mu\nu}, \\ E_3' &= F'^{03} = \Lambda_\mu^0 \Lambda_\nu^3 F^{\mu\nu}, \\ H_1' &= F'^{23} = \Lambda_\mu^3 \Lambda_\nu^3 F^{\mu\nu}, \\ H_2' &= F'^{31} = \Lambda_\mu^3 \Lambda_\nu^1 F^{\mu\nu}, \\ H_3' &= F'^{12} = \Lambda_\mu^1 \Lambda_\nu^2 F^{\mu\nu}. \end{aligned}\right\} \tag{6.39}$$

since

$$\Lambda_0^0 = \gamma, \quad \Lambda_1^0 = -\gamma v/c^3, \quad \Lambda_2^0 = \Lambda_3^0 = 0, \quad \Lambda_0^1 = -\gamma v, \quad \Lambda_1^1 = \gamma,$$

$$\Lambda_2^1 \Lambda_3^1 = 0, \quad \Lambda_0^2 = \Lambda_1^2 = \Lambda_3^2 = \Lambda_0^3 = \Lambda_1^3 = \Lambda_2^3 = 0, \quad \Lambda_2^2 = \Lambda_3^3 = 1,$$

then (6.39) becomes

$$\left.\begin{aligned} E_1' &= \gamma^2 E_1 - (\gamma^2 v^2/c^2) E_1 = E_1, \\ E_2' &= \gamma E_2 - (\gamma v/c^2) H_3, \\ E_3' &= \gamma E_3 + (\gamma v/c^2) H_3, \\ H_1' &= H_1, \\ H_2' &= \gamma H_2 - \gamma v E_3, \\ H_3' &= \gamma H_3 + \gamma v E_2. \end{aligned}\right\} \tag{6.40}$$

In the moving frame S′ there is both an electric and magnetic field. Thus moving electric fields produce magnetic fields; similarly moving magnetic fields produce electric fields. This is to be expected from Faraday's and Ampere's laws.

Further reading for Chapter 6

More extensive discussions are contained in Chapters 7, 8, and 9 of S. L. Anderson's *Principles of relativity physics*, Academic Press, New York (1967); in O. Costa de Beauregard's *Precis of special relativity*, Academic Press, New York (1966); and in Chapters 1–5 of C. Møller's *The theory of relativity*, Clarendon Press, Oxford (1955).

It is also succinctly described and extended to fluid motion in Chapter 2 of S. Weinberg's *Gravitation and cosmology*, Wiley, New York (1972).

7. The structure of special relativity

7.1. The Lorentz group

THIS chapter gives a taste of relativity physics with a modern flavour. It is self contained but would be best attempted by the student who has had some aquaintance with linear algebra, especially matrices and determinants and a little group theory. It is possible to turn to the final chapter for those who find this chapter difficult. As with the previous section the nature of the arguments is so typical in modern theoretical physics that even a summary reading may be of value.

In the previous chapter there was constant use of the condition of invariance or covariance under an arbitrary Lorentz transformation. The detailed form of such Lorentz transformations has not so far been determined, and indeed we saw in § 4.4 how complicated the transformation becomes even when the two related frames of reference have their axes parallel to each other. Yet the condition of covariance under the set of all Lorentz transformations is undoubtedly a very powerful one. It is not satisfied by Newtonian mechanics, for example, and we had to develop a new definition of force and of the manner in which forces, and more general quantities called tensors, transform under Lorentz transformations in order that a Lorentz covariant theory of dynamics could be developed.

At this stage it is necessary, therefore, to describe in some form or other the nature of an arbitrary Lorentz transformation. We will do this by looking at the set of all possible such transformations. This would appear to be taking a retrograde step, a collection of complicated objects being expected to have at least as much complexity as the individuals. Further reflection shows that need not be necessarily true; the properties of a class will be distinct from those of its elements. In any case it is the whole set of Lorentz transformations which is of greatest interest, and not any particular one, since the invariance or covariance of a quantity is defined as under the action of *all* Lorentz transformations. We turn, then, to discuss the totality of these latter objects.

In the previous chapter we found that any Lorentz transformation could be represented as a 4×4 matrix Λ which transforms a four-vector $\boldsymbol{x}$ to $\Lambda\boldsymbol{x}$. The further condition $\Lambda^{\mathrm{T}}g\Lambda = g$ was necessary in order that the velocity of light was invariant under such a transformation. We may thus define the set of all Lorentz transformations as the set $\mathscr{L}$ of all 4×4 matrices which satisfies (6.10),

$$\mathscr{L} = [\Lambda : \Lambda^{\mathrm{T}}g\Lambda = g] \tag{7.1}$$

(where $[x:P(x)]$ denotes the set of x with the property $P(x)$). Since each Lorentz transformation has 16 independent real matrix elements, $\mathscr{L}$ is a subset of 16-dimensional Euclidean space $\boldsymbol{R}^{16}$.

Let us now consider what possible combinations of Lorentz transformations can be obtained. We may combine any two such transformations Λ_1, Λ_2 so as to act one after the other,

$$\boldsymbol{x} \to \boldsymbol{x}' = \Lambda_1 \boldsymbol{x} \to \boldsymbol{x}'' = \Lambda_2 \boldsymbol{x}' = \Lambda_2 \Lambda_1 \boldsymbol{x}, \tag{7.2}$$

where $\Lambda_2\Lambda_1$ is the matrix product of the matrices Λ_2, Λ_1. This combination of Λ_1 followed by Λ_2 is also a Lorentz transformation, since it satisfies (6.10) if each of Λ_1, Λ_2 do,

$$(\Lambda_2\Lambda_1)^{\mathrm{T}} g \Lambda_2 \Lambda_1 = \Lambda_1^{\mathrm{T}} \Lambda_2^{\mathrm{T}} g \Lambda_2 \Lambda_1 = \Lambda_1^{\mathrm{T}} g \Lambda_1 = g.$$

We may thus multiply any number $\Lambda_1, \Lambda_2, \ldots, \Lambda_n$ of Lorentz transformations together, though we must remember that their order is important;

$$\Lambda_1 \Lambda_2 \neq \Lambda_2 \Lambda_1$$

in general. It is possible also to obtain the inverse of a Lorentz transformation Λ which undoes the action of Λ. This should be the inverse Λ^{-1} of Λ, since

$$\boldsymbol{x} \to \boldsymbol{x}' = \Lambda \boldsymbol{x} \to \Lambda^{-1} \boldsymbol{x}' = \Lambda^{-1} \Lambda \boldsymbol{x} = \boldsymbol{x}. \tag{7.3}$$

We have to show that Λ^{-1} is always a Lorentz transformation if Λ is. We first prove that Λ^{-1} exists as a matrix. By taking the determinant of both sides of (6.10) we have

$$(\det \Lambda)^2 (\det g) = \det g,$$

and since $\det g = -1$ then

$$(\det \Lambda)^2 = 1, \qquad \det \Lambda = \pm 1. \tag{7.4}$$

In particular, (7.4) shows that since $\det \Lambda \neq 0$, Λ has a matrix inverse Λ^{-1}, with $\Lambda^{-1}\Lambda = \Lambda\Lambda^{-1} = 1$, as we assumed in writing down (7.3).

Let us next obtain a variant of (6.10). For any $\Lambda \in \mathscr{L}$ ($x \in X$ denotes x belongs to the set X),

$$\Lambda g \Lambda^{\mathrm{T}} g \Lambda g \Lambda^{\mathrm{T}} = \Lambda g^3 \Lambda^{\mathrm{T}} = \Lambda g \Lambda^{\mathrm{T}}. \tag{7.5}$$

Since $\det (g\Lambda g \Lambda^{\mathrm{T}}) = (\det \Lambda)^2 (\det g)^2 = 1$ then $\Lambda g \Lambda^{\mathrm{T}}$ has an inverse. Thus we may divide both sides of (7.5) by the matrix $g\Lambda g\Lambda^{\mathrm{T}}$, and using that $g^2 = 1$, so $g^{-1} = g$, then

$$\Lambda g \Lambda^{\mathrm{T}} = g. \tag{7.6}$$

Comparing (7.6) with (7.1) we see that if $\Lambda \in \mathscr{L}$ then $\Lambda^{\mathrm{T}} \in \mathscr{L}$. We can finally show that Λ^{-1} is a Lorentz transformation if Λ is one, using the rule $(ABC)^{-1} =$

$C^{-1}B^{-1}A^{-1}$ for an inverse of products of matrices and that $g^{\mathrm{T}} = g$,

$$(\Lambda^{-1})^{\mathrm{T}} g \Lambda^{-1} = (\Lambda g \Lambda^{\mathrm{T}})^{-1}. \tag{7.7}$$

Using (7.6) and that $g^{-1} = g$ leads to

$$(\Lambda^{-1})^{\mathrm{T}} g \Lambda^{-1} = g, \tag{7.8}$$

so that $\Lambda^{-1} \in \mathscr{L}$.

We have shown that

(1) if $\Lambda_1, \Lambda_2 \in \mathscr{L}$ then $\Lambda_1 \Lambda_2 \in \mathscr{L}$;

(2) if $\Lambda \in \mathscr{L}$ then $\Lambda^{-1} \in \mathscr{L}$.

We may add to these two results a third result,

(3) $\mathbf{1} \in \mathscr{L}$,

since evidently the unit matrix satisfies (7.1).

Consider a set A of objects $a, b, c, \ldots$ for which there is a rule of binary combination $a, b \to (ab)$ which is associative, so that $((ab).c) = (a(bc))$, and which satisfies the condition that each element has an inverse a^{-1} and that there is an identity e in the set for which $ea = ae = a$ for any $a \in A$. Such a set of elements is called a group. Matrix multiplication is associative because the multiplication of real numbers which are the matrix elements is so. Therefore the set $\mathscr{L}$ of all Lorentz transformations constitutes a group under the binary combination of matrix multiplication. This group is called the Lorentz group. It is this which is at the heart of special relativistic dynamics.

7.2. The structure of the Lorentz group

The Lorentz group $\mathscr{L}$ can be represented as a certain subset of 16-dimensional real space. There are 16 separate conditions (7.1) on the matrix elements of any Lorentz transformation Λ, but only 10 of these are independent, since the matrices $\Lambda^{\mathrm{T}} g \Lambda$ and g are symmetric. Thus any $\Lambda \in \mathscr{L}$ is specified by six independent real parameters, and $\mathscr{L}$ will be a six-dimensional subset of $\boldsymbol{R}^{16}$. The condition (7.1) is quadratic in the matrix elements of Λ so that $\mathscr{L}$ might be expected to be quite a complicated subset of $\boldsymbol{R}^{16}$. There is a certain simplicity about it, however, which can be obtained from (7.1).

First of all (7.4) shows that every Lorentz transformation Λ has determinant equal to ± 1: $\det \Lambda = \pm 1$. Since the determinant of a matrix varies continuously as the matrix elements change we cannot move along a smooth curve in $\mathscr{L}$ to go from matrices with $\det \Lambda = +1$ to those with $\det \Lambda = -1$. In other words, the sets $\mathscr{L}_+ = [\Lambda : \det \Lambda = +1]$ and $\mathscr{L}_- = [\Lambda : \det \Lambda = -1]$ cannot be connected by any curve in $\mathscr{L}$; they are said to be disconnected. It is necessary to jump from $\mathscr{L}_+$ to $\mathscr{L}_-$ or vice versa by a discontinuous change of matrix elements.

It is possible to divide up $\mathscr{L}$ into further disconnected pieces by use of condition (7.1) more effectively. If we consider the (0, 0) component of (7.1) it is, in detail,

$$(\Lambda^{\mathrm{T}})^{\mu}_{0} g_{\mu\nu} \Lambda^{\nu}_{0} = g_{00}$$

or

$$(\Lambda^{0}_{0})^2 - \sum_{i=1}^{3} (\Lambda^{i}_{0})^2 = 1.$$

Thus

$$\Lambda^{0}_{0} = \pm \left(1 + \sum_{i=1}^{3} (\Lambda^{i}_{0})^2\right)^{\frac{1}{2}},$$

so that

$$\Lambda^{0}_{0} \geqslant +1, \quad \text{or} \quad \Lambda^{0}_{0} \leqslant -1. \tag{7.9}$$

Evidently the condition (7.9) divides $\mathscr{L}$ into two regions, which, as before, cannot be joined by a smooth curve; they are disconnected. The two regions are

$$\mathscr{L}^{\uparrow} = [\Lambda : \Lambda^{0}_{0} \geqslant +1], \qquad \mathscr{L}^{\downarrow} = [\Lambda : \Lambda^{0}_{0} \leqslant -1]. \tag{7.10}$$

The arrow in the definitions of $\mathscr{L}^{\uparrow}$ and $\mathscr{L}^{\downarrow}$ denote the effect of Λ on the time component of a four-vector. If $\Lambda \in \mathscr{L}^{\uparrow}$ then $\Lambda^{0}_{0} x^{0}$ has the same sign as x^{0}, whilst if $\Lambda \in \mathscr{L}^{\downarrow}$ then $\Lambda^{0}_{0} x^{0}$ has the opposite sign to x^{0}. Thus $\mathscr{L}^{\uparrow}$ is the set of Lorentz transformations which preserve the sign of time, and $\mathscr{L}^{\downarrow}$ is the set of those which change it.

In all, we may now divide $\mathscr{L}$ into four regions, $\mathscr{L}^{\uparrow}_{+}$, $\mathscr{L}^{\uparrow}_{-}$, $\mathscr{L}^{\downarrow}_{+}$, $\mathscr{L}^{\downarrow}_{-}$, with

$$\mathscr{L}^{\uparrow}_{+} = [\Lambda : \det \Lambda = +1, \Lambda^{0}_{0} \geqslant 1], \qquad \mathscr{L}^{\uparrow}_{-} = [\Lambda : \det \Lambda = -1, \Lambda^{0}_{0} \geqslant 1],$$

$$\mathscr{L}^{\downarrow}_{+} = [\Lambda : \det \Lambda = +1, \Lambda^{0}_{0} \leqslant -1], \qquad \mathscr{L}^{\downarrow}_{-} = [\Lambda : \det \Lambda = -1, \Lambda^{0}_{0} \leqslant -1].$$

These four regions are disconnected from each other, and are pictured in Fig. 21. The identity Lorentz transformation 1 is in $\mathscr{L}^{\uparrow}_{+}$, and because of this $\mathscr{L}^{\uparrow}_{+}$, is called the connected component of the identity.

In fact this splitting of $\mathscr{L}$ into four distinct pieces allows us to recognize two very important Lorentz transformations, that of parity P and time-reversal T. The parity operator changes the sign of space but not of time,

$$P : \boldsymbol{r} \to -\boldsymbol{r}, t \to t,$$

so has the matrix form

$$P = \begin{bmatrix} 1 & 0 & 0 & 0 \\ 0 & -1 & 0 & 0 \\ 0 & 0 & -1 & 0 \\ 0 & 0 & 0 & -1 \end{bmatrix}.$$

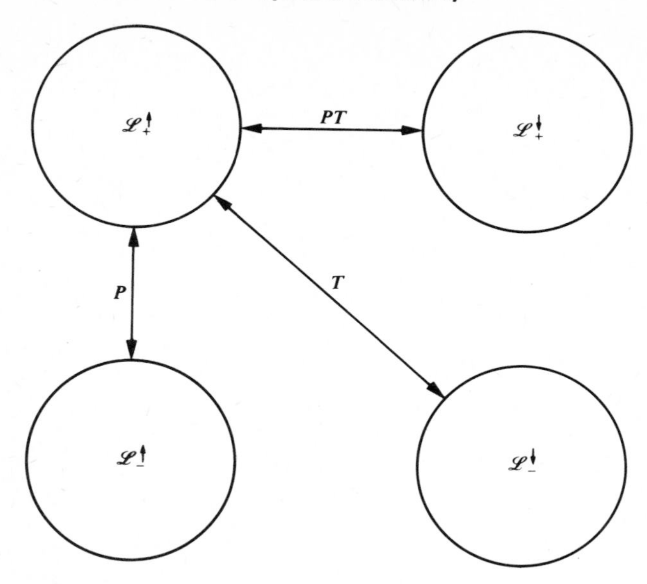

FIG. 21. The four disconnected portions $\mathscr{L}_+^\uparrow$, $\mathscr{L}_+^\downarrow$, $\mathscr{L}_-^\uparrow$, $\mathscr{L}_-^\downarrow$ of the Lorentz group, and the one-to-one relationship between them by means of the parity operator P and time-reversal operator T.

Evidently $\det P = -1$, $P_0^0 \geqslant -1$, so $P \in \mathscr{L}_-^\uparrow$. What is interesting is that if $\Lambda \in \mathscr{L}_+^\uparrow$ then $P\Lambda \in \mathscr{L}_-^\uparrow$, since

$$\det P\Lambda = \det P \,.\, \det \Lambda = -1, \qquad (P\Lambda)_0^0 = P_0^0 \Lambda_0^0 = \Lambda_0^0 \geqslant +1.$$

Also if $\Lambda \in \mathscr{L}_-^\uparrow$ then $P\Lambda \in \mathscr{L}_+^\uparrow$, since

$$\det P\Lambda = \det P \det \Lambda = +1, \qquad (P\Lambda)_0^0 = P_0^0 \Lambda_0^0 = \Lambda_0^0 \geqslant +1.$$

Thus the parity operator P allows any transformation Λ in $\mathscr{L}_+^\uparrow$ to be related to the corresponding transformation $P\Lambda$ in $\mathscr{L}_-^\uparrow$, and any transformation Λ in $\mathscr{L}_-^\uparrow$ is related to its partner $P\Lambda$ in $\mathscr{L}_+^\uparrow$. In other words, P allows a one-to-one correspondence to be set up between $\mathscr{L}_+^\uparrow$ and $\mathscr{L}_-^\uparrow$, as we have shown in Fig. 21.

The time-reversal operator T only changes the sign of time,

$$T : \boldsymbol{r} \to \boldsymbol{r}, t \to -t,$$

and so has matrix form

$$T = \begin{bmatrix} -1 & 0 & 0 & 0 \\ 0 & 1 & 0 & 0 \\ 0 & 0 & 1 & 0 \\ 0 & 0 & 0 & 1 \end{bmatrix}.$$

In this case, $\det T = -1$, $T^0_0 \leqslant -1$, so that $T \in \mathscr{L}^{\downarrow}_{-}$. As in the case of parity, T sets up a one-to-one correspondence, now between $\mathscr{L}^{\uparrow}_{+}$ and $\mathscr{L}^{\downarrow}_{-}$, again shown in Fig. 21. Finally, the operator PT is equal to -1, and has

$$\det PT = +1, \qquad (PT)^0_0 = -1.$$

Thus $PT \in \mathscr{L}^{\downarrow}_{+}$, and so it may be used to set up a one-to-one correspondence between $\mathscr{L}^{\uparrow}_{+}$ and $\mathscr{L}^{\downarrow}_{+}$.

We can conclude that it is not necessary to consider the whole of $\mathscr{L}^{\uparrow}_{+}$, $\mathscr{L}^{\downarrow}_{+}$, $\mathscr{L}^{\downarrow}_{-}$, $\mathscr{L}^{\uparrow}_{-}$ in detail, but only $\mathscr{L}^{\uparrow}_{+}$, P, and T. It is usual to call an element Λ of $\mathscr{L}^{\uparrow}_{+}$ a proper orthochronous Lorentz transformation, the proper referring to the condition $\det \Lambda = +1$. The distinction between proper and improper transformations and those which are orthochronous or not is important. No experimental results have yet been observed which are not invariant under proper Lorentz transformations, but parity violation in radioactive decays was discovered in 1957 on the suggestion of the Chinese physicists T. D. Lee and C. N. Yang, whilst time-reversal violation very likely occurs in related reactions. Thus we cannot ask for invariance of experimental results under the whole Lorentz group, but only the proper orthochronous Lorentz transformations constituting $\mathscr{L}_{+}$. Since these form a group in their own right, the product of any two of them being a transformation of the same sort, then the apparatus of group theory can be applied to $\mathscr{L}^{\uparrow}_{+}$. This has important implications for particle physics which we will come to shortly.

7.3. The rotation group

There are a particular set of Lorentz transformations which have everyday familiarity—the rotations in three-dimensional space. These involve no change of time, so have the matrix form

$$\Lambda = \begin{bmatrix} 1 & 0 & 0 & 0 \\ 0 & & & \\ 0 & & R & \\ 0 & & & \end{bmatrix}, \tag{7.11}$$

where R is some 3×3 matrix. The condition (7.1) now reads in the case of (7.11),

$$\Lambda^{\mathrm{T}} g \Lambda = \begin{bmatrix} 1 & 0 & 0 & 0 \\ 0 & & & \\ 0 & & R^{\mathrm{T}} & \\ 0 & & & \end{bmatrix} \begin{bmatrix} 1 & 0 & 0 & 0 \\ 0 & -1 & 0 & 0 \\ 0 & 0 & -1 & 0 \\ 0 & 0 & 0 & -1 \end{bmatrix} \begin{bmatrix} 1 & 0 & 0 & 0 \\ 0 & & & \\ 0 & & R & \\ 0 & & & \end{bmatrix}$$

$$= \begin{bmatrix} 1 & 0 & 0 & 0 \\ 0 & & & \\ 0 & & -R^{\mathrm{T}}R & \\ 0 & & & \end{bmatrix} = \begin{bmatrix} 1 & 0 & 0 & 0 \\ 0 & -1 & 0 & 0 \\ 0 & 0 & -1 & 0 \\ 0 & 0 & 0 & -1 \end{bmatrix}. \tag{7.12}$$

Evidently the $(0, 0)$, $(i, 0)$, and $(0, i)$ $(i = 1, 2, 3)$ components of (7.12) are satisfied, leaving the condition

$$R^{\mathrm{T}}R = 1. \tag{7.13}$$

This condition defines what is called an orthogonal matrix; the action of such a matrix on any three-vector $\boldsymbol{x}$ preserves the Euclidean length

$$x_1^2 + x_2^2 + x_3^2 = \boldsymbol{x}^{\mathrm{T}}\boldsymbol{x},$$

since if $\boldsymbol{x} \to \boldsymbol{y} = R\boldsymbol{x}$, then

$$\boldsymbol{y}^{\mathrm{T}}\boldsymbol{y} = \boldsymbol{x}^{\mathrm{T}}R^{\mathrm{T}}R\boldsymbol{x} = \boldsymbol{x}^{\mathrm{T}}\boldsymbol{x},$$

as required.

A length-preserving linear transformation may be a rotation about some axis. For example, a rotation through an angle of $-\theta$ about the z-axis can be considered as a rotation of the axes through θ about the z-axis. In that case we obtain (see Fig. 22), with OA $= x$, OB $=$ Y, OA$' = x'$, OB$' = y'$ that

$$\mathrm{ON} = x\cos\theta,\ \mathrm{NA}' = y\sin\theta,\ \mathrm{OA}' = x' = \mathrm{ON}+\mathrm{NA}' = x\cos\theta + y\sin\theta,$$

$$\mathrm{ON}' = y\cos\theta,\ \mathrm{BN}' = x\sin\theta,\ \mathrm{OB}' = y' = \mathrm{ON}'-\mathrm{B'N'} = y\cos\theta - x\sin\theta.$$

Introducing the unaltered component z, we have that the rotation through θ is represented by the rotation matrix

$$R(\theta) = \begin{bmatrix} \cos\theta & \sin\theta & 0 \\ -\sin\theta & \cos\theta & 0 \\ 0 & 0 & 1 \end{bmatrix}. \tag{7.14}$$

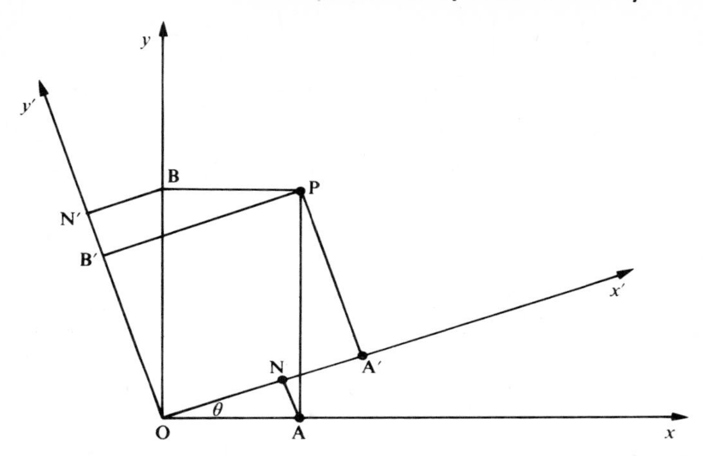

FIG. 22. The effect on the coordinate of a particle of a rotation of axes through an angle θ about the z-axis.

It is straightforward to verify that $R(\theta)$, given by (7.14), satisfies (7.13) and so is an orthogonal matrix.

As for the Lorentz group, we may consider the set of all orthogonal matrices, denoted by $O(3)$.

$$O(3) = [R : R^{\mathrm{T}}R = 1]. \tag{7.15}$$

It is not difficult to show that $O(3)$ is a group, the three-dimensional orthogonal or rotation group, since

(1) $1 \in O(3)$;
(2) if R_1, $R_2 \in O(3)$ then $(R_1R_2)^{\mathrm{T}}R_1R_2 = R_2^{\mathrm{T}}R_1^{\mathrm{T}}R_1R_2 = R_2^{\mathrm{T}}R_2 = 1$ by repeated use of (7.15);
(3) if $R^{\mathrm{T}}R = 1$ then $(RR^{\mathrm{T}})^2 = R(R^{\mathrm{T}}R)R^{\mathrm{T}} = RR^{\mathrm{T}}$, so $RR^{\mathrm{T}} = 1$.

Also if $R \in O(3)$ then $(\det R)^2 = 1$, $\det R \neq 0$, so R^{-1} exists; so does $(RR^{\mathrm{T}})^{-1}$, which we used implicitly in the previous step. Finally, we may prove $R^{-1} \in O(3)$ if $R \in O(3)$, since

$$(R^{-1})^{\mathrm{T}}R^{-1} = (RR^{\mathrm{T}})^{-1} = 1,$$

as required.

The group $O(3)$ has different connected parts in the same manner as the Lorentz group. Since $(\det R)^2 = 1$ then $\det R = +1$, so that $O(3)$ splits into two parts,

$$O_+(3) = [R : \det R = +1], \qquad O_-(3) = [R : \det R = -1]. \tag{7.16}$$

These two parts may be related by the parity operator

$$P = \begin{bmatrix} -1 & 0 & 0 \\ 0 & -1 & 0 \\ 0 & 0 & -1 \end{bmatrix},$$

since if $R \in O_+(3)$ then $PR \in O_-(3)$, and if $R' \in O_-(3)$ then $PR' \in O_+(3)$. Thus it is only necessary to consider either $O_+(3)$ or $O_-(3)$, and of the two $O_+(3)$ is the favoured since it is a group in its own right.

The conditions (7.15) are six in number, so that there are three independent matrix elements after these six restrictions have been used. These can be chosen to be an angle θ and a direction $\boldsymbol{n}$ in space (which only needs two angles to specify it). The corresponding matrix R represents a rotation through the angle θ about $\boldsymbol{n}$. This parametrization of R makes clear the rotational character of all the matrices.

The space of parameters $(\theta, \boldsymbol{n})$ can be pictured as points in a sphere of radius 2π, the point P with coordinates $(\theta, \boldsymbol{n})$ having radial distance OP equal to θ, the radius OP pointing along the direction $\boldsymbol{n}$. Since a rotation through the angle 2π about $\boldsymbol{n}$ will give the same result as of -2π about $\boldsymbol{n}$, it is necessary to identify points on the surface of this sphere which are at opposite ends of a diameter. This parameter-space is very important in further developments which involve the spin of the electron.

7.4. Representations of rotations

We may regard the group $O(3)$ as one composed of elements which are not necessarily matrices by 'abstract' elements for which we know only the group multiplication laws. It is then valuable to determine which sets of matrices can 'represent' these abstract group elements in a concrete fashion. This helps us to obtain copies of $O(3)$ which have other than 3×3 matrices as elements. Such a process of finding group representations is of great importance for all groups, but we will only restrict ourselves here to $O(3)$. This is not immediately relevant to special relativity, but the concept is of such fundamental importance in relativistic physics that its introduction in a non-relativistic frame may be of help.

We are interested, then, in a representation of $O(3)$; that is, for each element R in $O(3)$ we wish to find an $n \times n$ matrix $\Lambda(R)$, for some n, so that $\Lambda(R)$ has the multiplication law determined by that for R:

$$\Lambda(R_1)\Lambda(R_2) = \Lambda(R_1R_2)$$
$$\Lambda(R^{-1}) = \{\Lambda(R)\}^{-1}. \tag{7.17}$$

It is the condition (7.17) which determines if $\Lambda(R)$ defines a representation of

$O(3)$ or not. If (7.17) is valid for any R, R_1, $R_2 \in O(3)$ then $[\Lambda(R) : R \in O(3)]$ is said to form an n-dimensional representation of $O(3)$.

We always have a trivial representation, that of one dimension:

$$\Lambda(R) = 1, \quad \text{all} \quad R \in O(3), \tag{7.18}$$

which evidently satisfies (7.17). We can also immediately recognize the three-dimensional one:

$$\Lambda(R) = R, \quad \text{all} \quad R \in O(3). \tag{7.19}$$

We can obtain higher-dimensional representations by considering tensors similar to those in § 6.4. Following (6.24) we can define a two-index tensor T_{ij} as a set of nine quantities which transform under a rotation (using the summation convention only on indices taking values from 1 to 3),

$$T_{ij} \to T'_{ij} = R_{il} R_{jm} T_{lm}. \tag{7.20}$$

We see that if T_{ij} is symmetric, $T_{ij} = T_{ji}$, then this symmetry is preserved by (7.20),

$$T'_{ji} = R_{jl} R_{im} T_{lm} = R_{jm} R_{il} T_{ml} = R_{il} R_{jm} T_{lm} = T'_{ij}.$$

Furthermore, if the trace of T_{ij} is zero, the trace of T_{ij} being defined as T_{ii}, then so will be the trace of T'_{ij},

$$T'_{ii} = R_{il} R_{im} T_{lm} = (R^{\mathrm{T}} R)_{lm} T_{lm} = \delta_{lm} T_{lm} = T_{ll} = 0.$$

The set of components of a symmetric traceless two-index tensor T_{ij} are five in number, so (7.20) defines a new representation on them of dimension 5. We can see that any tensor $\bar{T}_{ij}$ can be expressed first as

$$\bar{T}_{ij} = \tfrac{1}{2}(\bar{T}_{ij} + \bar{T}_{ji}) + \tfrac{1}{2}(\bar{T}_{ij} - \bar{T}_{ji}), \tag{7.21}$$

and then the first term $\bar{T}_{ij}$ on the right-hand side of (7.21) can be written as

$$\bar{T}_{ij} = \tfrac{1}{2}(\bar{T}_{ij} + \bar{T}_{ji}) = (\bar{T}_{ij} - \tfrac{1}{3}\delta_{ij}\bar{T}_{ll}) + \tfrac{1}{3}\delta_{ij}\bar{T}_{ll}. \tag{7.22}$$

The first term on the right-hand side of (7.22) is a symmetric traceless tensor, so has five independent components, while the second term has one component, the trace of $\bar{T}_{ij}$, and gives the one-dimensional representation. Thus the decomposition of (7.21) and (7.22) expresses any two-index tensor as the sum of a single component quantity, proportional to the identity matrix, a set of three independent quantities combined in an anti-symmetric tensor $\tfrac{1}{2}(\bar{T}_{ij} - \bar{T}_{ji})$, and a set of five independent components of a symmetric traceless tensor. Numerically, these dimensions of the corresponding representations add to nine,

$$9 = 1 + 3 + 5, \tag{7.23}$$

as they should. This decomposition is as far as it is possible to go; there are no further subsets of two-index tensors which preserve their identity under the representation (7.20). Representations which have no subsets of tensors preserving their identity are called "irreducible'.

The representation (7.20) is called the product representation of two three-dimensional ones, and if each of these representations is denoted by **3** then the product of them is $\mathbf{3}\times\mathbf{3}$. The dimensional equation (7.23) now becomes the rule for the decomposition of the product representation $\mathbf{3}\times\mathbf{3}$ into its irreducible components:

$$\mathbf{3}\times\mathbf{3} = \mathbf{1}+\mathbf{3}+\mathbf{5}. \tag{7.24}$$

It is possible to consider higher tensors $T_{i_1\ldots i_n}$ transforming as

$$T_{i_1\ldots i_n} \to T'_{i_1\ldots i_n} = R_{i_1 j_1}\ldots R_{i_n j_n} T_{j_1\ldots j_n}. \tag{7.25}$$

The representation (7.25) is evidently the product representation $\mathbf{3}\times\mathbf{3}\times\cdots\times\mathbf{3}$, the product being taken n times. This product may be decomposed into its irreducible components to obtain an extension of (7.24); we will not consider this further here.

Analysis of representations of $O(3)$ are of great importance for defining the spin of a particle. It is always possible to transform to the rest frame of a massive particle; the remaining transformations which should not change the description of the particle are then rotations of space. It is these rotations, and particularly their representations, which are crucial to distinguish one particle type from another. It turns out that all known particles have states which can be regarded as the tensors of some irreducible representation $O(3)$. If the dimension of the representation is $(2n+1)$ then the particle spin is proportional to n the constant of proportionality being $\hbar$, Planck's constant h divided by 2π. It can also be shown that, besides the representations with $n = 0, 1, 2, \ldots$ with dimensions 1, 3, 5, ... , which we have considered here, there are also those with $n = \frac{1}{2}, \frac{3}{2}, \ldots$ of dimension 2, 4, ... ; the first of these represents the electron, neutrino, mu-meson, proton, and neutron, and many other of the excited companions of the last two.

7.5. Elementary particle scattering

One of the most important uses of Lorentz invariance in modern physics is in the study of the dynamics of the interactions between elementary particles. These particles travel at speeds close to that of light, and so have to be described in a manner which includes the various features of close-to-light travel we have already found. The most natural way to do this is to incorporate Lorentz invariance into the dynamical equations. We have already briefly mentioned how the rotation and Lorentz groups are important in characterizing the spin of the various elementary particles. In this section we will consider how the energy and momentum variables of particles are to be treated in a Lorentz-invariant scattering problem.

For simplicity we will consider a process involving only particles of mass m and concentrate on the elastic scattering of one such particle from another (with no other particles being produced). Let us suppose that the energy-momentum four vectors of these two particles are initially p_1 and p_2; after their interaction these variables are found to be p_3 and p_4 respectively. The process of scattering shown in Fig. 23 can be represented by a scattering

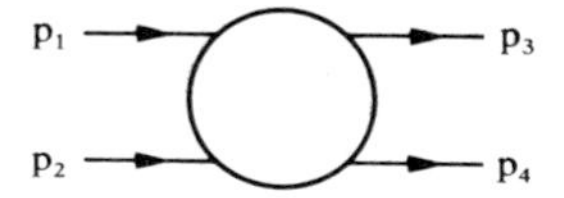

FIG. 23. The symbolic representation of the elastic scattering of two identical particles, with energy-momentum four-vectors which are initially p_1, p_2 and finally p_3, p_4.

amplitude $f(p_1, p_2; p_3, p_4)$. This latter would describe the proportion of particles which start with initial energy-momentum four-vectors p_1 and p_2 to end up after scattering from each other with these four-vectors having values p_3 and p_4.

As a first step in incorporating Lorentz invariance we note that the laws of conservation of energy and momentum can be written as

$$p_1 + p_2 = p_3 + p_4 \tag{7.26}$$

Therefore the scattering amplitude is only a function of three of the four-vectors p_1, p_2, p_3, p_4. However this still makes twelve variables in all, so that further reduction of variable is evidently desirable.

This can be achieved by using the condition that each of the four-vectors entering in (7.26) describes a particle of mass m, so that if we use the notation

$$p_2 = p_\mu p^\mu$$

then

$$p_1^2 = p_2^2 = p_3^2 = p_4^2 = m^2. \tag{7.27}$$

There are four equations in (7.27) each of which we may solve for the energy component:

$$p_{io} = +\sqrt{\boldsymbol{p}_i^2 + m^2} \qquad (i = 1, 2, 3, 4) \tag{7.28}$$

where $p_i = (p_{io}, \boldsymbol{p}_i)$ and the positive square root is chosen on the r.h.s. of (7.28) since energy is positive. We thus obtain a further reduction of variables by the use of (7.28), with only 8 independent variables remaining.

It is possible to obtain a considerable diminution of this set of variables by requiring that the scattering be Lorentz invariant. This means that as observed in a Lorentz-transformed frame of reference S the scattering is the same as that in the original frame S (with the same origin). Otherwise scattering results would enable a differentiation to be made between one frame and another.

If Λ is the homogeneous Lorentz transformation from S to S′ then the condition of Lorentz invariance of scattering is expressed as

$$f(p_1, p_2, p_3, p_4) = f(\Lambda p_1, \Lambda p_2, \Lambda p_3, \Lambda p_4) \tag{7.29}$$

It is necessary to solve (7.29) in a manner which reduces the number of variables in the scattering amplitude. One solution can be seen immediately: f is a function only of the independent Lorentz invariants which can be constructed from the four vectors p_1, p_2, p_3, and p_4. The largest set of such invariants is the set of sixteen quantities $p_i p_j$ where i and j take the values from one to four. Since $p_i p_j = p_j p_i$ then only 10 of these quantities are independent. We may reduce the number of invariants by energy-momentum conservation (7.26) to six by replacing p_4 by $p_1 \rightarrow p_2 - p_3$: these invariants would then be

$$p_1^2, p_2^2, p_3^2, p_1 p_2, p_1 p_3, p_2 p_3 \tag{7.30}$$

where we have the restriction (7.27):

$$p_4^2 = (p_1 + p_2 - p_3)^2 = m^2 \tag{7.31}$$

Using (7.27) again in (7.31) this can be rewritten, on evaluating the quadratic expression, as

$$p_1 p_2 - p_1 p_3 - p_2 p_3 + 2m^2 = 0 \tag{7.32}$$

Since also $p_1^2 = p_2^2 = p_3^2 = m^2$ there are thus only two independent invariants, say $p_1 p_2$ and $p_1 p_3$ in the set (7.30).

These two invariants may be given a direct physical interpretation if we consider the explicit components of the variables p_1, p_2, p_3, p_4 in the centre of mass system for which the total momentum of the incoming particles and of the outgoing ones is zero:

$$\boldsymbol{p}_1 + \boldsymbol{p}_2 = \boldsymbol{p}_3 + \boldsymbol{p}_4 = O.$$

Then

$$p_1 = (p_0, \boldsymbol{p}), p_2 = (p_0, -\boldsymbol{p}), p_3 = (p_0', \boldsymbol{p}'), p_4 = (p_0', -\boldsymbol{p}')$$

where

$$p_0 = \sqrt{\boldsymbol{p}^2 + m^2}, \qquad p_0' = \sqrt{\boldsymbol{p}'^2 + m^2}.$$

From energy conservation

$$2p_0 = 2p_0'$$

so that

$$\boldsymbol{p}^2 = \boldsymbol{p}'^2.$$

Finally the invariants p_1p_2 and p_1p_3 are:

$$p_1p_2 = m^2+2\boldsymbol{p}^2, \qquad p_1p_3 = \boldsymbol{p}^2+m^2-\boldsymbol{p}^2\cos\theta$$

where θ is the angle between $\boldsymbol{p}$ and $\boldsymbol{p}'$:

$$p^2\cos\theta = \boldsymbol{p}\,.\,\boldsymbol{p}'.$$

Clearly the two variables of interest are $\cos\theta$ and p_0, and f can be expressed as a function $F(\cos\theta, p_0)$, θ being the scattering angle and p_0 the energy of each of the incoming particles.

The result indicates that when particles of a given energy scatter from each other in a Lorentz invariant fashion the scattering distribution can only depend on the scattering angle. This property of elastic scattering actually follows if only rotation invariance is required instead of the full Lorentz invariance. However Lorentz invariance indicates which combinations of variables are important. Indeed the invariants:

$$s = (p_1+p_2)^2, \qquad t = (p_1-p_3)^2 \tag{7.33}$$

are the ones most naturally occurring, s being the square of the total energy of the two particles and t being the invariant momentum transfer; in the centre of mass:

$$s = (p_{10}+p_{20})^2 = 4p_0^2, \qquad t = -(p_1-p_3)^2 = -2\boldsymbol{p}^2(1-\cos\theta).$$

It is in terms of these variables that simple expressions for the scattering amplitudes have been attempted.

Together with the very important reduction in numbers of variables in the case of many particle processes, Lorentz invariance has proved of inestimable value in helping to comprehend the variety of processes occurring for the elementary particles.

Further reading for Chapter 7

Groups, including the rotation group, are covered by H. Margenau and G. Murphy, in Chapter 15 of *The mathematics of physics and chemistry*, Van Nostrand, New York (1962).

Tensors are described succinctly in B. Spain's *Tensor calculus*, Oliver and Boyd, Edinburgh (1965).

The Lorentz group is also considered by R. F. Streater and A. Wightman, from the point of view of Chapter 7, in Chapter 1 of *PCT and all that*, Benjamin, New York (1964).

8. Extensions of special relativity

8.1. Faster than light?

THIS is a question which has caused much speculation, especially in the realm of science fiction. Recently it has come into the scientific fold. We will discuss here some of the recent developments.

The factor $\gamma(v) = (1-v^2/c^2)^2$ present in the Lorentz transformation becomes imaginary if $v > c$. This has often been used to argue that it would never be possible to travel faster than light, since space and time would then become imaginary. In the framework of Lorentz transformations it is quite clear that one cannot make any sense of a frame of reference travelling faster than light with respect to another reference frame. Since it is evident that frames of reference can travel slower than light there cannot therefore be any which move faster.

This result has been used in the past to conclude that nothing can ever travel faster than light. It is possible to emphasize this by considering the increase of the mass and energy of a particle with its velocity. It was shown graphically in Fig. 18 (p. 50) how the mass of the particle increases indefinitely as v approaches c, there becoming infinite. The amount of energy available to any external agent trying to accelerate the particle is presumably finite, so that the velocity of light presents a barrier to all faster motion. Even if there were an infinite amount of energy available, such as that from the rest energies of all the particles in an infinite universe, it would still not be enough to crash through the light barrier, since once $v > c$ the value of the energy $\gamma(v)m_0c^2$ will become imaginary. The light barrier appears to be a very effective one indeed, being an infinitely high wall at c with absolutely nothing on the other side.

Yet accepting that picture involves giving in too readily. For a particle to be worthy of the name it would seem to be necessary for it to possess observable (so real) energy and momentum. These are required to vary proportionally to the factor $\gamma(v)$, thus providing an extension of the expressions (5.19) to values $v > c$. But the constant of proportionality, the rest mass m_0, may be chosen to be purely imaginary: $m_0 = \mathrm{i}\mu_0$ with μ_0 a positive number. In terms of μ_0 we may write the energy E and momentum $\boldsymbol{p}$, which are now real, of such a particle as

$$E = \mu_0 c^2\left(\frac{v^2}{c^2}-1\right)^{-\frac{1}{2}}, \qquad \boldsymbol{p} = \mu_0 \boldsymbol{v}\left(\frac{v^2}{c^2}-1\right)^{-\frac{1}{2}}. \tag{8.1}$$

Evidently E and $|\boldsymbol{p}|$ both decrease as v increases; their behaviour is shown in Fig. 23. We note that the energy decreases as the particle speeds up, finally

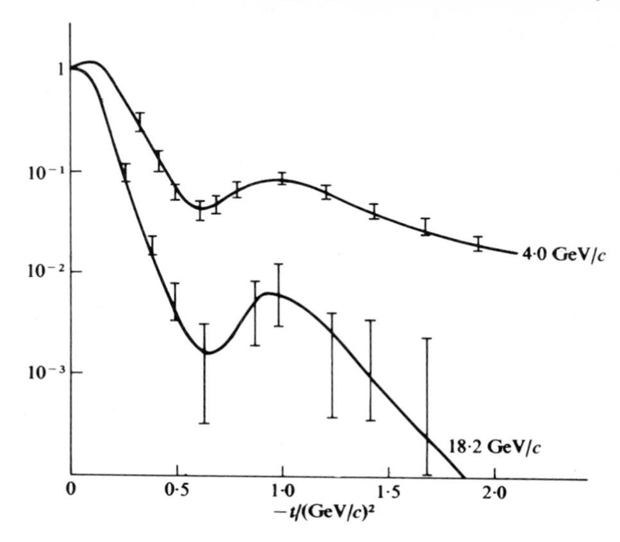

FIG. 24. The proportion of mesons of momentum 4 and 18·2 GeV/c (in the laboratory frame) scattered from protons; the reaction also includes charge exchange.

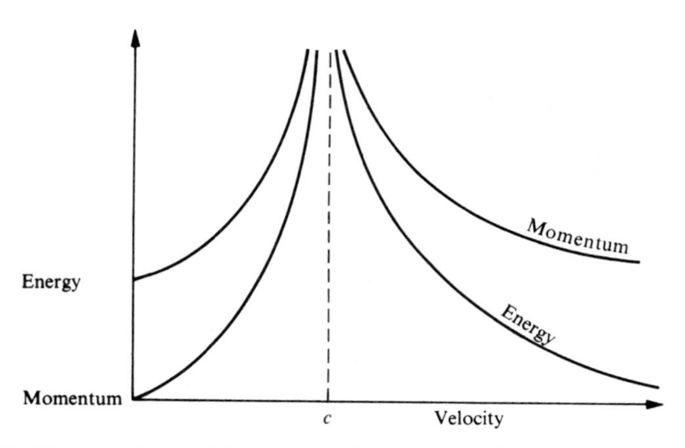

FIG. 25. The dependence of the energy and momentum of a particle on its velocity for faster than light particles with imaginary rest mass (tachyons) and normal particles travelling slower than light (tardons).

becoming zero when the velocity is infinite; the momentum decreases to the constant value $\mu_0 c$ as $|\boldsymbol{v}|$ tends to infinity. This is just the opposite to the change of E and $\boldsymbol{p}$ for a particle moving slower than light, as we see from Fig. 25.

Such particles moving faster than light have been called tachyons, from the Greek *tachys*, meaning swift. A tachyon always travels faster than light; the light barrier stops its decrease of speed. This is just the reverse of the effect of the barrier on particles moving slower than light; these particles are sometimes appropriately called tardons. Provided tardons and tachyons never have their identities confused there should be no difficulty of imaginary energy or momentum; nor of imaginary rest mass for tachyons, provided they always travel faster than light.

If tachyons exist they present a new difficulty. Suppose such a particle, travelling with velocity $u > c$, is emitted from the point P and received at Q, at time Δt later, as measured in the frame of reference S in which P and Q are both at rest. If the coordinate separation of P and Q is Δx then $u = \Delta x/\Delta t$. If $\Delta t'$, $\Delta x'$ are the values, measured in a frame S′ moving with velocity v with respect to S, of the time and space separations of the emission and reception of the tachyon, then

$$\Delta t' = \gamma(v)\left(\Delta t - v\Delta\frac{x}{c^2}\right), \ \Delta x' = \gamma(v)(\Delta x - v\Delta t).$$

Thus

$$\Delta t' = \gamma(v)\Delta t\left(1 - \frac{uv}{c^2}\right), \tag{8.2}$$

and $\Delta t'$ can be made negative if $v > c^2/u$. Since c^2/u is less than c it is possible to choose an infinity of frames S′ for which $\Delta t' < 0$. At the same time the energy E' of the tachyon, as observed in S′, is related to the energy E and momentum $\boldsymbol{p}$ in S by

$$E' = \gamma(v)\left(E - \frac{vp}{c^2}\right) = \gamma(v)E\left(1 - \frac{uv}{c^2}\right). \tag{8.3}$$

(Since the tachyon velocity $u = p/E$). The factor on the right-hand side of (8.3) is identical to that on the right-hand side of (8.2); if $\Delta t' < 0$ then $E' < 0$. Negative-energy could be made as negative as desired by taking v large enough they would be expected to be produced copiously, even from a vacuum. Free space would become completely full of them.

It is possible to avoid both of these difficulties by the 'reinterpretation principle'. By this, the emission of a negative-energy tachyon is reinterpreted as the absorption of one of positive energy, and conversely the absorption of a negative-energy tachyon is to be considered as the emission of one of positive energy. Such a principle embodies exactly what would occur during observations of these particles if only positive energies were allowable. For loss of negative energy would be seen as the increase of energy, gain of negative energy as the loss of energy. We show in Fig. 26 this reinterpretation principle at work; the reinterpreted tachyon travels with positive energy from its earlier time of emission to its later absorption.

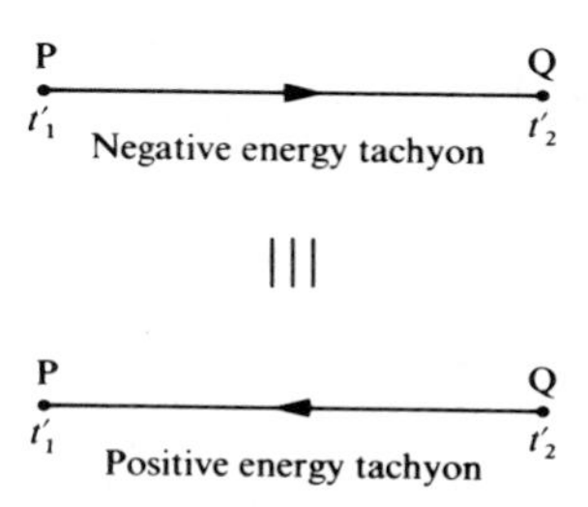

FIG. 26. The reinterpretation principle for tachyons at work; the negative energy tachyon emitted from P at a time after it is absorbed at Q is reinterpreted as a positive energy tachyon absorbed by P after it had been emitted from Q.

By this means causality—cause always precedes effect—is saved simultaneously with positivity of energy. But violation of causality can still occur in spite of the reinterpretation principle. In Fig. 27 we present the space–time diagram of an observer A signalling into his own past. He sends out a tachyon, taken as travelling with infinite speed, which is received by his colleague B who is travelling away from him with velocity v. B then immediately sends an infinitely fast tachyon, as measured in B's frame S′, back to A.

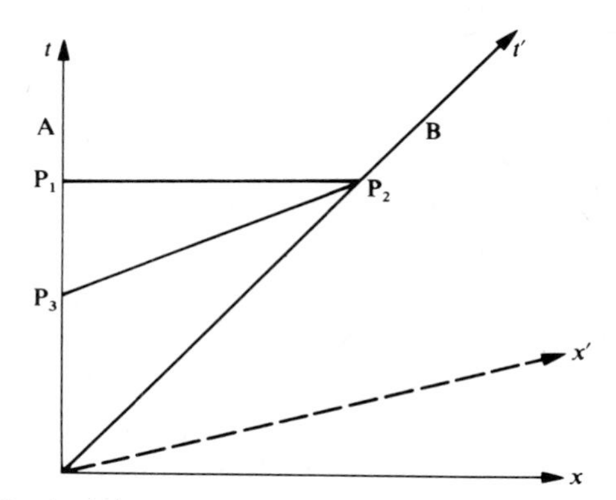

FIG. 27. Signalling back into one's past. Observer A sends an infinitely fast tachyon to the moving observer B, who then sends a similar one back to A. This latter returns to A in his past.

As in Fig. 27 let P_1 be the event of A sending the first tachyon, with coordinate labels $(0, t_A)$ in A's frame S. Since the world line of B is the line $x = vt$ in S then the event P_2 of B receiving A's tachyon will have coordinates (vt_A, t_A) in S. The coordinates of P_2 in S′ are then

$$(0, \gamma(v)\{t_A - (v^2 t_A/c^2)\}) = (0, t_A(1 - v^2/c^2)^{\frac{1}{2}}).$$

We can now calculate the coordinates of the event P_3 when A receives the return tachyon from B. In S′, A moves along the line $x' = -vt'$. In S′ the events P_2 and P_3 are related by an instantaneous tachyon, so have the same time coordinate. Therefore P_3 has labels $(-t_A v(1 - v^2/c^2)^{\frac{1}{2}}, t_A(1 - v^2/c^2)^{\frac{1}{2}})$ in S′ which in S are

$$\left(0, \gamma(v)\left\{t_A\left(1 - \frac{v^2}{c^2}\right)^{\frac{1}{2}} - t_A\left(\frac{v^2}{c^2}\right)\left(1 - \frac{v^2}{c^2}\right)^{\frac{1}{2}}\right\}\right) = \left(0, t_A\left(1 - \frac{v^2}{c^2}\right)\right).$$

This means that A can signal back into his past by an amount $(v^2/c^2)t_A$, which can be as large as desired if t_A is chosen large enough.

The reinterpretation principle does not remove this possibility of time travel into one's past. The principle would lead to the tachyon from A appearing to B as if he himself had sent it, and A thinking *he* had transmitted B's tachyon instead of receiving it. But this altered picture of the nature of the events does not alter their causal order; as far as A is concerned he sends out a tachyon to B which 'causes' A to send out another tachyon at an *earlier* time. It is this feature of A's having had to have performed an act before the action which caused the first act which is a true paradox. For example, A could have signalled back into his past to have his parents prevented from ever meeting, and so made his own conception impossible.

In spite of this very difficult paradox, tachyons have been carefully searched for. The simplest way that tachyons could interact with ordinary matter would be if the tachyon were charged. Pairs of oppositely charged tachyons might then be created by light of sufficiently high intensity. Photons of about 700 keV arising from the decay of cesium 104 were used in an experiment at Princeton University by T. Alväger and M. N. Kriesler. They tried to detect any tachyons produced in a lead shield surrounding the decaying material by looking for the Čerenkov radiation the charged tachyons should emit. This radiation is the shock wave emitted by a particle travelling faster than light in a medium; it is the radiation equivalent of the sonic boom. None was seen, nor have any of various other experiments produced any hint of the interaction of tachyons with matter.

This absence of tachyons is all to the good, as far as causality is concerned. It has been shown more recently that the violation of causality would occur at a much deeper level if such particles existed. Even so it is necessary to keep an open mind on this, since the degree of violation may be so small as to produce only minimal macroscopic effects. Tachyons may yet be discovered.

8.2. Acceleration and special relativity

We saw in Chapter 5 how to develop laws of particle dynamics which are in agreement with the principles of special relativity, namely, the constancy of the velocity of light and the absence of any preferred inertial frame (Lorentz covariance of the dynamical laws). It was shown in § 6.4 how this programme could be pushed through for electrically charged matter interacting through the electromagnetic field. However, this was an easy case, since the constancy of the velocity of light is inherent in Maxwell's equation of electrodynamics. It is far more difficult to develop a theory of gravity which agrees with the principles of special relativity because the Newtonian force of gravity is an instantaneous one. To be in concordance with special relativity the effects of gravity cannot spread at a speed faster than that of light. This requires a new theory altogether, and in particular it requires a Lorentz covariant field theory which will need to have close similarity to Maxwell's equations.

Another problem which has been avoided so far is the restriction of all discussion to frames of reference in uniform motion relative to each other. This neglect of accelerated motion is disturbing, and evidently it would be preferable to construct a dynamical theory of events in space and time in which the labels assigned by *any* choice of frame of reference are irrelevant to observable results. This idea of formulating dynamical laws in a coordinate independent fashion is that of general relativity, by means of which Einstein formulated his celebrated theory of gravitation in 1915. This was achieved by his recognizing clearly that constant acceleration has the same effects as a uniform gravitational field; acceleration and gravity are two aspects of the same phenomenon.

It is not possible here to develop the general theory of relativity or Einstein's theory of gravity, but it would be relevant to consider the first steps and some of the related experimental results. We will start by considering the effect of a uniform acceleration on an inertial frame of reference S in the absence of forces.

Suppose a particle P is at the point with coordinate x in S (we will only discuss one space dimension for simplicity). Then since there is no force on P the coordinate x will satisfy (in Newtonian mechanics)

$$\frac{\mathrm{d}^2x}{\mathrm{d}t^2} = 0.$$

If the frame S is now accelerated to the right with constant acceleration a the position x' of the particle, as seen in S, will now be $x' = x - \frac{1}{2}at^2$, so that

$$\frac{\mathrm{d}^2x'}{\mathrm{d}t^2} = -a.$$

If the particle has inertial mass m then this latter equation can be written as the Newtonian equation

$$\frac{m\,\mathrm{d}^2x'}{\mathrm{d}t^2} = -ma. \tag{8.4}$$

Thus the particle appears as if it is being acted on by a constant force $-ma$, the net acceleration $-a$ being independent of the mass of the particle.

This independence of motion on mass is exactly the same as has been found for particles moving under gravity on the earth's surface. Neglecting the variation of the acceleration g due to gravity with position a particle moves under the Newtonian law

$$\frac{m\,\mathrm{d}^2x}{\mathrm{d}t^2} = -m_g g, \tag{8.5}$$

where the mass m_g is the passive gravitational mass of the particle. This quantity determines how strongly a particle is acted on by a given gravitational field. It need not equal the inertial mass m entering on the left-hand side of (8.5), so that in general the value of the acceleration of the particle resulting from (8.5) will depend on the particle as well as the size of g.

Careful experiments have shown that there is no such dependence of acceleration on particle type; for all particles

$$m = m_g. \tag{8.6}$$

This was shown to be true to 1 part in 10^8 by Baron Eötvös in the last century, and since then improved to 1 part in 10^{11} by Dicke and his colleagues at Princeton University. This result shows that (8.4) and (8.5) are identical, so that the effects of a constant acceleration are the same as those of a constant gravitational field. Thus an observer furnished only with an accelerometer to observe his surroundings would not be able to distinguish, when his instrument indicated he was undergoing constant acceleration, whether he was either (a) in free space accelerating relative to the fixed stars or (b) at rest on the surface of a gravitating body. Similarly, if his accelerometer read zero he would be unable to distinguish if he was either (a) uniformly moving in outer space relative to the stars or (b) freely falling in the earth's gravitational field, so undergoing acceleration relative to the stars.

This impossibility of distinguishing between linear acceleration and constant gravity has been elevated to the status of a principle, called the principle of equivalence. It is on this principle that Einstein based his theory of gravitation. We will only use this principle here to derive a formula for the effect of a gravitational field on the frequency of light—the gravitational red shift.

Consider two observers A and B at rest in a gravitational field, say on the surface of the earth. We suppose A is at a height l above B. Then the effect of the field will be identical to that caused by a constant acceleration g

upwards. A pulse of light is emitted from B, in the direction of A, with frequency ν_B. In the meantime A has experienced an acceleration lasting a time l/c, so he has acquired a velocity $v = gl/c$ away from B. Thus by the Doppler effect (§ 4.2) the frequency of the light seen by A will not be ν_B but

$$\nu_A = \nu_B\left(1-\frac{v}{c}\right) = \nu_B\left(1-\frac{gl}{c^2}\right).$$

In rising in the gravitational field the light has had its frequency reduced by amount gl/c^2; it has suffered a red shift. Similarly if the light had fallen in the gravitational field it would have increased its frequency and been correspondingly blue-shifted.

That a gravitational red shift occurs is known from the spectra of massive O-type stars, which may be 100 times heavier than the sun and about 5 times as large across. The expected red shift is then about 3×10^{-5}, or, multiplying by c, equal to about $10\,\text{km s}^{-1}$ of the velocity of light. This is in very good agreement with the values given in Table 5 (taken from the talk of R. J. Trumpler at the Conference on the Jubilee of Relativity Theory held in Basle in 1956). The effect has also been observed on earth by R. V. Pound and G. A. Rebka with an l of 22 m. The predicted frequency shift was $2{\cdot}46\times10^{-15}$, in excellent agreement with the experimental result of $(2{\cdot}57\pm0{\cdot}26)\times10^{-15}$; recently this agreement has been improved to 1 per cent.

TABLE 5

Gravitational red shifts of O-type stars

Cluster	Star	Red shift/km s^{-1}
JC1805	2	$12{\cdot}4\pm2{\cdot}2$
	3	$2{\cdot}8\pm2{\cdot}0$
JC 1848	1	$10{\cdot}4\pm4{\cdot}8$
	10	$4{\cdot}6\pm5{\cdot}4$
	2	$6{\cdot}2\pm3{\cdot}5$
NGC 2544	9	$6{\cdot}8\pm1{\cdot}3$
	15	$13{\cdot}6\pm1{\cdot}7$
	8	$6{\cdot}4\pm1{\cdot}6$
NGC 2264	1	$9{\cdot}8\pm1{\cdot}2$
NGC 2353	1	$16{\cdot}1\pm1{\cdot}6$
NGC 6231	50	$16{\cdot}4\pm2{\cdot}6$
NGC 6604	1	$13{\cdot}6\pm4{\cdot}1$
NGC 6611	1	$9{\cdot}0\pm2{\cdot}1$
	2	$4{\cdot}1\pm3{\cdot}7$
	3	$9{\cdot}9\pm2{\cdot}4$
NGC 6823	1	$11{\cdot}6\pm3{\cdot}4$
	2	$7{\cdot}7\pm3{\cdot}6$
NGC 6871	5	$15{\cdot}6\pm1{\cdot}6$

8.3. Whither relativity?

It is natural to conclude this book with a brief discussion of where present problems lie in the theory of relativity. It is clear that special relativity is a very successful theory, at least as far as particles travelling at high speeds are concerned. The validation of Einstein's equation of mass with energy, '$E = mc^2$', has been completely achieved in nuclear reactions, whilst the increase of energy with velocity and the phenomenon of time dilation show perfect agreement between Einstein's theory of special relativity and the experimental results.

There are much more subtle effects which have caused considerable trouble recently. They arise from the attempt to formulate a Lorentz covariant theory of wave mechanics. This latter is the formalism which has to replace that of Newtonian mechanics when the probabilistic nature of matter is realized. For example, an electron in a circular orbit round a nucleus inside an atom does not always have the same distance from the nucleus. Its orbit can only be described by a wave function $\psi(\boldsymbol{r}, t)$ whose squared modulus $|\psi(\boldsymbol{r}, t)|^2\, \mathrm{d}^3\boldsymbol{r}$ gives the probability of finding the electron in the volume element $\mathrm{d}^3\boldsymbol{r}$ around $\boldsymbol{r}$ at time t. This probability distribution $|\psi|^2$ is shown for a typical case in Fig. 28; whilst the electron spends the majority of its time near r_0 it is not there with certainty.

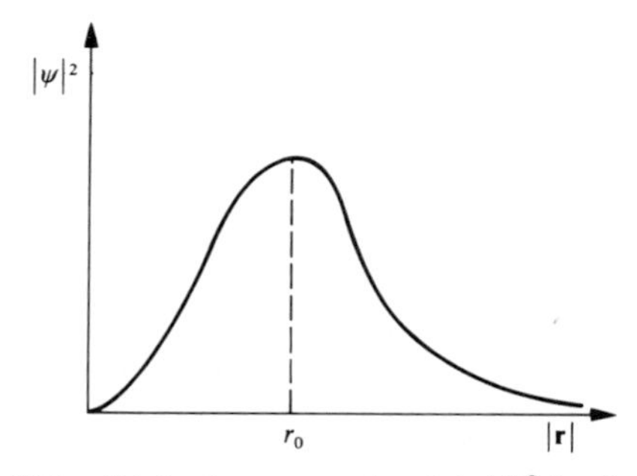

FIG. 28. The probability distribution proportional to $|\psi|^2$ for finding an electron at a certain distance from the nucleus in the hydrogen atom, according to wave mechanics. The distance r_0 is that of maximum probability.

This loss of certainty in wave mechanics is compensated for by the remarkable explanation it gives of atomic and nuclear processes. But it is initially a non-relativistic theory. Wave mechanics was introduced in about 1925. Since then many attempts have been made to marry it with special relativity. All of these were plagued with what are known as the 'ultraviolet divergences', which were infinite quantities which entered in any physical quantity being calculated. In 1948 it was shown that these divergences could be removed by redefinition of the physical quantities for the particles—their masses and charges. This renormalization programme, as it is called, was eminently successful. Among the predictions were small changes of the magnetic moments and energy levels of the electron and other charged particles; these were found to be in complete agreement with experiment, and remain so till today.

The renormalization programme initially suffered from having explicit removal of the ultraviolet divergences, with expressions such as $\infty - \infty$ having to be taken as 0. This was evidently highly unsatisfactory, but since then the theory has been formulated so that such ambiguous expressions never enter; the successful predictions remain the same. In such a manner wave mechanics has been fruitfully joined to special relativity. There are still unsolved questions concerning the detailed mathematical structure of such theories, even such basic questions as whether theories of this sort exist at all. But recent results are beginning to show that the theories indeed have the mathematical properties they were expected to possess on physical grounds.

There is still the great difficulty of extending this marriage of relativity and wave mechanics to general relativity, that is, to gravity. It has proved impossible so far to obtain a wave-mechanical theory of gravity which can be renormalized. This is a pressing problem, especially so since the realization of the possible existence of black holes in space—very massive stars which have collapsed in on themselves after have used up all their nuclear fuel. In the centre of such an object matter would be annihilated to nothing very rapidly, if Einstein's theory of general relativity were to be believed. Thus the theories break down completely here; they cannot cope with such a situation. It is hoped that wave mechanics will save the day. It appears now that it will only do so if Einstein's theory is suitably modified so as to be renormalizable. The most exciting problem of the present, and undoubtedly for quite a few years to come, is in attempting to bring wave mechanics and general relativity fruitfully together; their union will bless us with great understanding of the Universe.

Further reading for Chapter 8

See J. G. Taylor, Faster than light?, *Sci. Jl* **5A**, 42 (1969), and for deeper difficulties with faster-than-light travel, J. G. Taylor and M. M. Broido. Does Lorentz invariance imply causality?, *Phys. Rev.* **174**, 1606 (1968).

On gravitation and general relativity see Einstein's book mentioned earlier or the useful introduction by C. W. Kilmister, *General theory of relativity*, Pergamon Press, Oxford (1973), as well as the previously mentioned texts by Anderson, Møller, and Weinberg (see Further reading in Chapters 3 and 6).

The present difficulties caused by black holes are discussed by J. G. Taylor in *Black holes: the end of the universe?*, Souvenir Press, London (1973) and more technically in J. G. Taylor, Gravitational collapse and black holes, *Phys. Bull.* **24**, 654–6 (1973). It is also considered more extensively by C. Misner, K. Thorne, and J. Wheeler, in *Gravitation*, Freeman, London (1973) (especially Part VII).

Appendix

AN $n \times n$ matrix is a square array

$$A = [a_{ij}] = \begin{bmatrix} a_{11} & a_{12} & \dots & a_{1n} \\ a_{21} & a_{22} & \dots & a_{2n} \\ \dots & \dots & \dots & \dots \\ \dots & \dots & \dots & \dots \\ a_{n1} & a_{n2} & \dots & a_{nn} \end{bmatrix}.$$

The (i,j)th matrix element a_{ij}, sometimes denoted by $(A)_{ij}$, will be, for us, a real number. It is in the ith row and the jth column of the matrix array A. It is possible to add the matrices $A = [a_{ij}]$ and $B = [b_{ij}]$ by adding the corresponding matrix elements,

$$C = A+B, \quad \text{where} \quad c_{ij} = a_{ij}+b_{ij},$$

and to multiply any matrix A by the real number λ by multiplying each matrix element of A by λ,

$$C = \lambda A, \quad \text{where} \quad c_{ij} = \lambda a_{ij}.$$

Two $n \times n$ matrices A, B may also be multiplied together,

$$C = AB, \quad \text{where} \quad c_{ij} = a_{ik}b_{kj} \tag{A.1}$$

(the summation convention of summing a repeated index over its values from 1 to n is being used, so that $a_{ik}b_{kj}$ is, when written out in full, the sum $\sum_{k=1}^{n} a_{ik}b_{kj}$). The transpose A^{T} of a matrix A is defined by interchanging the rows and columns,

$$(A^{\mathrm{T}})_{ij} = a_{ji}.$$

A is said to be symmetric if $A^{\mathrm{T}} = A$ or $a_{ij} = a_{ji}$.

It is also useful to define a product between an $n \times n$ matrix and a column vector $\boldsymbol{x}$ with n components,

$$\boldsymbol{y} = A\boldsymbol{x}, \quad \text{where} \quad y_i = a_{ij}x_j. \tag{A.2}$$

Then if $\boldsymbol{y} = B\boldsymbol{x}$, $\boldsymbol{z} = A\boldsymbol{y}$, we have $\boldsymbol{z} = AB\boldsymbol{x}$, where AB is the product of the two matrices defined above in (A.1).

Matrices arise naturally in the representation of system of equations such as in (A.2) in a compact form. We may also solve such equations in a matrix form; if we can define the inverse A^{-1} of the matrix A then $\boldsymbol{y} = A\boldsymbol{x}$ has solution $\boldsymbol{x} = A^{-1}\boldsymbol{y}$. The problem is thus to obtain A^{-1}. To do that we turn to the determinant det A of A, defined by

$$\det A = \sum_{\rho} \varepsilon_{\rho} a_{1\rho(1)} a_{2\rho(2)} \dots a_{n\rho(n)}. \tag{A.3}$$

In (A.3) the summation is over all permutations ρ of $1 \dots n$ and ε_{ρ} is the signature of ρ, being $(-1)^{n_{\rho}}$, n_{ρ} being the number of transpositions (interchanges of two numbers) in going from $\rho(1), \dots, \rho(n)$ to the natural order $1, \dots, n$. We also define the cofactor A_{ij} to a_{ij} as being $(-1)^{i+j}$ times the determinant of the $(n-1) \times (n-1)$ matrix obtained by erasing the ith row and jth column in A.

We may write, by concentrating on the first factor in each term of (A.3),

$$\det A = \sum_{j=1}^{n} a_{ij} A_{ij} = \sum_{j=1}^{n} a_{ij} A_{ij}, \tag{A.4}$$

where i is any of $(1, \dots, n)$ in (A.4). Also when $i \neq k$

$$\sum_{j=1}^{n} a_{ij} A_{kj} = 0, \tag{A.5}$$

the left-hand side of (A.5) being a determinant with the ith and kth rows equal, so being zero (as may be seen directly from the definition of (A.3), the terms cancelling in pairs). We now define the inverse matrix A^{-1} to A by

$$(A^{-1})_{ij} = (\det A)^{-1} A_{ji}. \tag{A.6}$$

This is correct, since from (A.6)

$$\begin{aligned}(AA^{-1})_{ij} = a_{ik}(A^{-1})_{kj} = (\det A)^{-1} a_{ik} A_{jk} &= 1 \qquad (i = j)\\ &= 0 \qquad (i \neq j).\end{aligned}$$

Then $AA^{-1} = I$, where I is the $n \times n$ unit matrix. Evidently (A.6) only exists if det $A \neq 0$, as was used in the text.

We note also that the definition (A.3) of the determinant results in the property

$$\det AB = \det A \,.\, \det B.$$

For we may write

$$\det AB = \sum_{\rho} \varepsilon_{\rho} a_{1i_1} b_{i_1\rho(1)} a_{2i_2} b_{i_2\rho(2)} \dots a_{ni_n} b_{i_n\rho(n)}, \tag{A.7}$$

and if $(1, 2, \dots, n) \rightarrow (i_1, \dots, i_n)$ is the permutation τ we may write the right-hand side of (A.7) as

$$\sum_{\rho,\tau} \varepsilon_{\rho} a_{1\tau(1)} b_{\tau(1)\rho(1)} a_{2\tau(2)} b_{\tau(2)\rho(2)} \dots a_{n\tau(n)} b_{\tau(n)\rho(n)}. \tag{A.8}$$

We may write $\rho = \tau\sigma$, where σ is another permutation of $1, \ldots, n$, and regard τ and σ as the independent variables in (A.8); using $\varepsilon_\rho = \varepsilon_{\tau\sigma} = \varepsilon_\tau\varepsilon_\sigma$, (A.8) becomes

$$\sum_\tau \varepsilon_\tau a_{1\tau(1)} \ldots a_{n\tau(n)} \sum_\sigma \varepsilon_\sigma b_{\tau(1)\tau\sigma(1)} \ldots b_{\tau(n)\tau\sigma(n)}. \tag{A.9}$$

The second factor in (A.9) is det B, any re-ordering of the numbers $1, \ldots, n$ on the right-hand side of (A.3) clearly only permuting the terms in that sum amongst each other but not changing the value of the sum. Since this second factor is independent of τ the first factor of (A.9) may now be summed over τ to give the further term det A, so proving that (A.9) is equal to det A . det B, so proving (A.7).

Finally, we prove that if det $A \neq 0$, det $B \neq 0$, so that A^{-1}, B^{-1}, $(AB)^{-1}$ all exist, then $(AB)^{-1} = B^{-1}A^{-1}$, for $B^{-1}A^{-1}$ certainly acts as the inverse to AB,

$$B^{-1}A^{-1}AB = B^{-1}B = 1 = ABB^{-1}A^{-1}.$$

Since the inverse to AB is unique it must thus be equal to $B^{-1}A^{-1}$. This reversal of ordering also occurs in taking the transpose of a product

$$(AB)^{\mathrm{T}} = B^{\mathrm{T}}A^{\mathrm{T}},$$

since

$$\{(AB)^{\mathrm{T}}\}_{ij} = (AB)_{ji} = a_{jk}b_{ki} = b_{ki}a_{jk} = (B^{\mathrm{T}})_{ik}(A^{\mathrm{T}})_{kj} = (B^{\mathrm{T}}A^{\mathrm{T}})_{ij}.$$

Problems

Problems for Chapter 2

2.1. The head of SPECTRE has devised a fiendish plot to liquidate his mortal enemy 007; he will cause a toxic rain to fall on the deserted valley where the intelligence agent is staying with a companion. Fortunately Mr. Bond's latest bag of tricks contains an umbrella. If the rain falls at a speed of $10\ \mathrm{m\ s^{-1}}$, at what angle with the vertical must 007 hold the umbrella to stay dry as he and his companion run at a speed of 10 m.p.h. to catch the next bus out of the deserted valley?

2.2. (a) Show that, if the true time interval between successive eclipses of Jupiter's moons is τ (equal to the orbital period of a moon), the observed time interval is $\tau+\Delta\tau$, where $\Delta\tau \sim R\Delta\theta \sin\theta/c$ and $\Delta\theta \sim V\tau/R$. Hence show that the accumulated time lag as the earth moves from closest to farthest approach to Jupiter (ignoring Jupiter's own change of position) is $2R/c$.
(b) The period of Jupiter's second moon, Europa, is 3·55 days. What is the maximum discrepancy between this true period and the time interval between successive eclipses as observed at the earth? What is the accumulated time lag between A and B? (Earth's orbital speed $= 30\ \mathrm{km\ s^{-1}}$; its orbital radius $= 1{\cdot}49 \times 10^{8}$ km.)

2.3. The ether-wind theory of the Michelson–Morley experiment is discussed in the text for the special case where the arms of the interferometer are parallel and perpendicular to the wind. Consider the general case for an angular setting θ. Prove that, for equal arms of length l, the time difference for the two paths is given to a good approximation by

$$\Delta t \sim v^{2} l \cos 2\theta/c^{3}.$$

2.4. In one arm of a Michelson interferometer there is placed a closed tube of length 0·2 m with transparent end pieces through which the light passes freely. A fringe pattern is observed using light of wavelength $5{\cdot}9 \times 10^{-7}$ m, and then the air is evacuated out of the tube. By how many fringes will the fringe pattern shift? The speed of light in air is $\{1-(2{\cdot}9\times 10^{-4})\}c$.

Problems for Chapter 3

3.1. Frame S′ has a speed $v = 0{\cdot}6c$ relative to S. Clocks are adjusted so that $t = t' = 0$ at $x = x' = 0$.
(a) An event occurs in S at $t = 2\times 10^{-7}$ s at a point for which $x = 50$ m. At what time does the event occur in S′?
(b) If a second event occurs at (10 m, 3×10^{-7} s) in S, what is the time interval between the events as measured in S′?

3.2. The inertial coordinate systems S and S′ move with speed $\frac{1}{2}c$ with respect to each other. Draw a space–time diagram relating these two systems. (Let the axes of x and ct for S be at right angles in your drawing.)

3.3. Two events occur at $(x_1, t_1) = (2\text{ m}, 10^{-8}\text{ s})$, $(x_2, t_2) = (30\text{ m}, 10^{-7}\text{ s})$ in a fixed frame S. Calculate the coordinates of these events in a new frame S′ moving at $0{\cdot}6c$ with respect to S, and show that $(x_1 - x_2)^2 - c^2(t_2 - t_1)^2$ is the same as $(x_1' - x_2')^2 - c^2(t_2' - t_1')^2$, giving its value.

3.4. Two events in a frame S are

$$E_1 : x_1 = x_0, t_1 = x_0/c; \qquad E_2 = x_2 = 2x_0, t_2 = x_0/2c.$$

Find the velocity of the frame in which E_1 and E_2 are simultaneous and their spatial separation in this new frame. How could it have been calculated directly from the coordinates of E_1 and E_2 in S?

3.5. A rocket, of proper length l_0, travels at constant velocity relative to a frame S. When the nose A′ of the ship passes a given point A, at $t = t' = 0$, a light signal is sent from A′ to the tail B′ of the ship. Find
(a) the ship time t' when the signal reaches the tail;
(b) the time t_1 (in S) when the signal reaches the tail;
(c) the time t_2 (in S) when the tail of the ship passess A.

3.6. A spaceman B travels at velocity v with respect to his control centre A. Both can send infinitely fast messages (in their own frames). A sends such a signal at time t to B, who then sends the message at infinite speed back to A. Show, by algebraic calculation using the appropriate Lorentz transformations, how much earlier A receives B's signal than his (A's) original transmission.

3.7. Show, by direct calculation, that the result of two successive Lorentz transformations,

$$(x, t) \to (\gamma_1(x - v_1 t), \gamma_1(t - v_1 x/c^2)) = (x', t'),$$

$$(x', t') \to (\gamma_2(x' - v_2 t'), \gamma_2(t' - v_2 x'/c^2) = (x'', t''),$$

is also a Lorentz transformation, and find the velocity ($\gamma_j = (1 - v_j^2/c^2)^{-\frac{1}{2}}$.)

Problems for Chapter 4

4.1. Two reference frames S and S′ move with speed $v(<c)$ with respect to each other. An object moves with velocity $\mathbf{u}$ in S and $\mathbf{u}'$ in S′. If $v = \frac{3}{4}c$, and $\mathbf{u}'$ has components $u_x' = -2c$, $u_y' = u_z' = 0$, show that the components of $\mathbf{u}$ are $u_x = \frac{5}{2}c$, $u_y = u_z = 0$. How can you account for the fact that u_x and u_x' have opposite signs? Experiment (on paper) with some other examples in which either or both of the combining velocities is greater than c. Is there any physical significance to such situations?

4.2. Consider three galaxies A, B, and C. An observer in A measures the velocities of C and B and finds they are moving in opposite directions each with a speed of $0{\cdot}7c$ relative to him. Thus, according to measurements in his frame, the distance between them is increasing at the rate $1{\cdot}4c$. What is the speed of A observed in B? What is the speed of C observed in B?

4.3. Measurements in two frames, S and S′, are related by the usual Lorentz transformations, with $v = 0{\cdot}6c$. At $t' = 10^{-7}$ s, a particle leaves the point $x' = 10$ m, travelling in the x'-direction with a constant velocity u' equal to $-\frac{1}{3}c$. It is brought to rest suddenly at $t' = 3 \times 10^{-7}$ s (all measurements in S′). As measured in S:
(a) What was the velocity of the particle during its trip?
(b) How far did it travel?

4.4. Consider two inertial frames, S and S′, related in the usual manner. A light signal of frequency ν_0 in S is emitted from the point $x = -l$ at time $t = -l/c$. What is the frequency of the light signal in S′?

4.5. A and B are twins. A goes on a trip to α-Centauri (4 light-years away) and back again. He travels at speed 0·6*c* with respect to the earth both ways, and transmits a radio signal every 0·01 years in his frame. His twin B similarly sends a signal every 0·01 years in his own rest frame.
(a) How many signals emitted by A before he turns around does B receive?
(b) How many signals does A receive before he turns around?
(c) What is the total number of signals each twin receives from the other?
(d) Who is younger at the end of the trip, and by how much? Show that the twins both agree on this result.

Problems for Chapter 5

5.1. How fast must an electron move in order for its energy to be (1) 10^3 times (2) 10^6 times greater than its rest mass?

5.2. A particle as observed in a certain reference frame has a total energy of 5 GeV and a momentum of 3 GeV/*c* (that is, *cp*, which has the dimension of energy, is equal to 3 GeV).
(a) What is its energy in a frame in which its momentum is equal to 4 GeV/*c*?
(b) What is its rest energy in MeV?
(c) What is the relative velocity of the two reference frames?

5.3. A particle of rest mass m_0 and kinetic energy $2m_0c^2$ strikes and sticks to a stationary particle of rest mass $2m_0$. Find the rest mass M_0 of the composite particle.

5.4. A 'photon rocket' uses pure radiation as the propellant. If the initial and final rest masses of the rocket are M_i and M_f show that the final velocity of the rocket relative to its initial rest frame is given by the equation

$$M_i/M_f = \{(c+v)/(c-v)\}^{\frac{1}{2}}.$$

5.5. An atom in an excited state of energy Q_0 above the ground state moves toward a scintillation counter with speed v. The atom decays to its ground state by emitting a photon of energy Q (as recorded by the counter), coming completely to rest as it does so. If the rest mass of the atom is m, show that

$$Q = Q_0\{1+(Q_0/2mc^2)\}.$$

5.6. The neutral pi-meson (π^0) decays into two γ-rays (and nothing else). If a π^0 (whose rest mass is 135 MeV) is moving with a kinetic energy of 1 GeV:
(a) What are the energies of the γ-rays if the decay process causes them to be emitted in opposite directions along the pion's original line of motion?
(b) What angle is formed between the two γ-rays if they are emitted at equal angles to the direction of the pion's motion?

Problems for Chapter 6

6.1. Two inertial frames, S and S′, move with speed v with respect to one another. Along the x-axis of S lies an infinitely long wire which is composed of stationary positive charges and negative charges moving in the x-direction with speed v.

Thus a current flows through the wire, although the net charge density in S is everywhere zero. What is the net charge density in S′? Does this result imply that total charge is not conserved in a Lorentz transformation? Explain. (See D. L. Webster, *Am. J. Phys.* **29**, 841 (1961) for a discussion and explanation of this phenomenon.)

Problems for Chapter 7

7.1. Write down the inverse of the matrix describing the Lorentz transformation with velocity along the x-axis, and interpret it physically.

7.2. By using the product rule for matrices show that the product of two successive rotations through angles θ and ϕ about the z-axis is a rotation through the angle $(\theta+\phi)$ about the z-axis.

7.3. Show that a rotation through an angle θ about the x-axis can be written, for small θ, as $(1+\theta L_x)$, where

$$L_x = \begin{bmatrix} 0 & 0 & 0 \\ 0 & 0 & 1 \\ 0 & -1 & 0 \end{bmatrix},$$

and write down similar expressions for small y- and z-rotations (where $\sin\theta \sim 0$ $\cos\theta \sim 1$ for small θ). Use this to prove that the difference between successive small rotations about the x- and y-axes and about the y- and x-axes is a small rotation about the z-axis, and find the angle of rotation resulting.

7.4. A tensor T_{ij} is defined to transform under the rotation matrix R, as

$$T'_{ij} = R_{il}R_{jm}T_{lm}.$$

Prove that such a set of nine quantities T_{ij} gives a representation of the rotation matrices, in that the transformation of T_{ij} under the two successive rotation matrices R and S is the same as that under the product SR.

Solutions to problems

Chapter 2

2.1. $24°$.

2.2. (b) About 30 s; about 16 min.

2.4. $3{\cdot}3 \times 10^{-3}$ Å ($3{\cdot}3 \times 10^{-4}$ nm).

Chapter 3

3.1. (a) $1{\cdot}25 \times 10^{-7}$ s; (b) $2{\cdot}25 \times 10^{-7}$ s.

3.2. (c) $x' = 0{\cdot}58$, $ct' = 0{\cdot}58$, $x = 1{\cdot}73$, $ct = 1{\cdot}73$.

3.3. ($0{\cdot}25$ m, $0{\cdot}75 \times 10^{-8}$ s); (15 m, 5×10^{-8} s); $-(55)^2$ m^2.

3.4. $-0{\cdot}5c$; $\frac{1}{2}\sqrt{3}x_0$.

3.5. (a) l_0/c; (b) $(l_0/c)\{(c-v)/(c+v)\}^{\frac{1}{2}}$.

3.6. $(v^2/c^2)t$

3.7. $(v_1+v_2)/(1+v_1v_2/c^2)$.

Chapter 4

4.2. $0{\cdot}7c$; $0{\cdot}94c$.

4.3. (a) $\frac{1}{3}c$; (b) 20 m.

4.4. (a) $v_0\{(1-\beta)/(1+\beta)\}^{\frac{1}{2}}$; (b) $v_0(1-\beta)$; (d) 2×10^{-15}.

4.5. (a) 533 (b) 267 (c) A gets 1333, B gets 1067 (d) A is younger than B by 2 years 8 months.

Chapter 5

5.1. (i) $0{\cdot}9999995$; (ii) $0{\cdot}9999999999995$.

5.2. (a) $5{\cdot}66$ GeV (b) $4{\cdot}04$ GeV (c) $0{\cdot}187c$.

5.3. $\sqrt{17}m_0$.

5.6. (a) $1{\cdot}131$ GeV, 4 MeV; (b) about $14°$.

Chapter 6

6.1. $\beta(1-\beta^2)^{-\frac{1}{2}}/c$. Total charge is conserved, but one must consider the complete circuit, not the charge density along a limited section.

Chapter 7

7.1.

$$\begin{array}{c} \\ x \\ y \\ z \\ t \end{array}\begin{array}{c} \begin{array}{cccc} x & y & z & t \end{array} \\ \begin{bmatrix} \gamma & 0 & 0 & -\gamma v \\ 0 & 1 & 0 & 0 \\ 0 & 0 & 1 & 0 \\ -\gamma v/c^2 & 0 & 0 & \gamma \end{bmatrix} \end{array}$$

7.2.

$$\begin{bmatrix} \cos\theta & \sin\theta & 0 \\ -\sin\theta & \cos\theta & 0 \\ 0 & 0 & 1 \end{bmatrix}$$

7.3. θ_x, θ_y.

Index